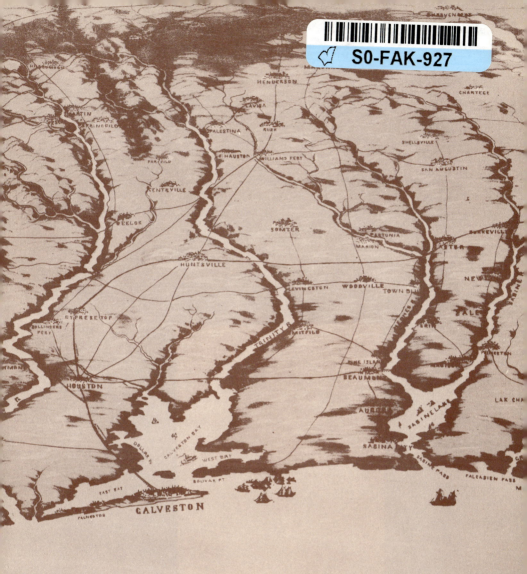

*O F M E X I C O*

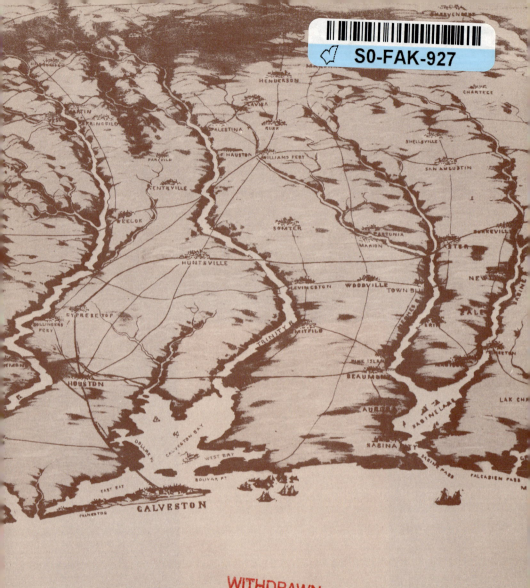

O F   M E X I C O

LETTERS FROM A TEXAS SHEEP RANCH

UNIVERSITY OF ILLINOIS PRESS, URBANA, 1959

# LETTERS
# FROM A TEXAS
# SHEEP
# RANCH

Written in the years 1860 and 1867 by

George Wilkins Kendall to Henry Stephens Randall

EDITED BY HARRY JAMES BROWN

# PREFACE

The letters here presented which were written in 1860 were discovered a few years ago at Elm Place, the home of William G. Markham (1836–1916) near Avon, New York. They were in a box with several hundred other letters, most of which were to or from Henry S. Randall, dealing with sheep and wool matters. These letters, which have been cited in the notes as the Randall Papers, are now in the Collection of Regional History, Cornell University.

William G. Markham was for many years one of the most active and best known sheep breeders in the United States. For more than twenty years he was secretary of the National Wool Growers' Association. Whether the Randall correspondence was turned over to him by Randall himself or by a member of the latter's family after Randall's death in 1876 is not known. Miss Linda Puffer and Mrs. Henry Selden, niece and daughter respectively of Mr. Markham, made the Randall Papers available to me. For this, for their interest in the project, and for their many kindnesses during my visits to Elm Place, I am very grateful.

The letters which were written in 1867 are owned by the New-York Historical Society, and I wish to thank the Director of that institution, Mr. R. W. G. Vail, for permission to use them.

I am also indebted to Mrs. Edith M. Fox, the indefatigable Curator of the Collection of Regional History, Cornell University, without whose help the letters at Elm Place would not now have come to light; to Mr. Lewis Randall, Lansdowne, Pennsylvania, for the portrait of Henry S. Randall; to Professor Fayette Copeland of the University of Oklahoma, author of *Kendall of the Picayune*, for information; to Professor Vernon Carstensen of the University of Wisconsin, who encouraged me to prepare the letters for book publication; to Mr. Wayne Rasmussen and Mr. Charles Burmeister of the United States Department of Agriculture, for reading and commenting on the manuscript; to the New Orleans *Times-Picayune* for a photograph of the portrait of George Wilkins Kendall by Thomas Hicks; to the Library of Congress for the map used as end papers and for other services; to Mrs. Donald Elder of the Cortland County (New York) Historical Society, for courtesies during my visits to Cortland; and to the Library of the North Texas State Teachers College at Denton, for the loan of a microfilm copy of the *Texas Almanac;* to John Houdek for assistance in proofreading and indexing.

The project was completed during a sabbatical year for which I owe thanks to Michigan State University.

# CONTENTS

There are two callings, friend Randall, in this world, which imperatively demand *practical* knowledge, viz:—navigation and sheep raising. In fair weather all goes well with theory, but when storms and foul weather come, then a practical man is needed at the helm. You may make a voyage to Calcutta and back, and watch every rope and every movement on board from daylight until dark, and be up of night as well: you may think you thoroughly understand navigation: but the first gale, cyclone, hurricane or tempest that comes, where are you? Gone! So with navigating a flock of sheep through the voyage of a year in Texas: you must serve your time "after the flock," as "before the mast," in order to insure a safe termination.

GEORGE WILKINS KENDALL

# INTRODUCTION

The steamer *Louisiana,* from New Orleans, docked at Powderhorn on Matagorda Bay in late April, 1856, discharging among its passengers George Wilkins Kendall, age forty-six, his "elegant and amiable" French wife some twenty years his junior, her sister Henriette, two of his four children, Georgina, nearly six, and the sickly three-year-old William, and a German governess for the children. For all of them except the head of the household, Texas was a new and strange, even a fearful land.

The next day the family started for the interior in a high-wheeled wagon drawn by two horses. In their wake came a load of trunks hauled by four mules, and a big load of furniture to which twelve yoke of oxen were hitched. Included among the household effects were a Chickering piano from Boston, and many items which had come from France—a red velvet sofa, a red velvet rug, bedroom pieces, delicate china and other things which were to add a note of refinement to the crudities of western Texas.[1]

Slowly the family moved toward its destination—a ranch in the hills a few miles above the German settlement at New Braunfels. Rains added to the discomforts of the journey. At a day's end the weary travelers were not sure of finding decent accommodations for the night. Mornings they were up

---

[1] New Orleans *Daily Picayune,* 27 April 1856 (letter from G.W.K.); Fayette Copeland, *Kendall of the Picayune* (Norman, 1943), 279–80.

early and on the way again. At many houses along the road
they noted packs of "cross, barking worthless curs." [2]

At length they reached their new home, a green expanse of
valley and hills on the upper Guadalupe. The family quarters
consisted of a house with a gallery running around it, with
four rooms downstairs (and a cellar below), and an attic
divided into bedrooms by burlap, and a kitchen separate
from the house. Not until a family had moved out could Mrs.
Kendall go to work in earnest. In time her cherished posses-
sions were in order in this strange setting and the music of
the Boston piano was heard on the frontier.[3]

Kendall started out to visit his sheep, which were pastured
some thirty miles westward at Post Oak Spring, near Boerne.
He laughed when one of his shepherds approached him,
struck by the incongruity between the stereotype of the
gentle shepherd with crook and lute, and the vision before
him. He wrote: "You may judge of my surprise when a Fra
Diavolish looking fellow stalked up to me, a double-barreled
gun on his shoulder, a Bowie knife hanging at one side, and
one of Colt's six-shooters on the other, at the same time
announcing from a mouth completely hidden behind a fierce
surrounding of beard and mustache that he was keeper of
the flock! A more brigandish looking shepherd was surely
never seen; yet the arsenal he carried about him he deemed
necessary for his own protection against Indians." Real frontier country, this Texas, where pioneering qualities of high
order were required for survival.[4]

A few weeks later he was in the metropolis of the region,
San Antonio, and concluded after observing the goings-on
of a day, that the town "was a place." Camels belonging to
the army came to town, scattering the horses. News of the
nomination of Buchanan and Breckinridge caused the lager
beer in the saloons to flow freely. A load of ice arrived from

[2] *Daily Picayune*, 22 May 1856 (letter from G.W.K.).

[3] Copeland, *Kendall*, 281.

[4] *Daily Picayune*, 22 May 1856 (letter from G.W.K.).

Indianola, making it possible to have your brandy with ice water if you were willing to pay five cents extra. Reports came in of the killing of a dozen or so Comanches near Fort Chadbourne.[5]

For Kendall the move to Texas was the fruition of a long dream. Many years before, when he was working in a printing office, a fellow worker had heard him "descant on his ambition to own a frontier farm with plenty of room about it for his dog and his gun." [6] But between the early dream and that April when he first brought his family to Texas, things had happened which had made his life seem like that of a hero of a novel of derring-do, and had made him one of the best known journalists of his day.

Born in Mount Vernon, New Hampshire, in 1809, George spent his first seven years with his parents, who moved frequently, living in a number of Vermont towns and in Montreal. Then he went back to New Hampshire to live at Amherst with his maternal grandfather, the jovial Deacon Wilkins, for ten years. In the same town another boy destined to win fame in journalism was also growing up—Horace Greeley. At sixteen Kendall worked as an apprentice for the short-lived Amherst *Herald,* and then went to Boston where he became an apprentice for a time on the *Statesman.* He went on to New York but was soon at Albany, facing westward. During the next six or seven years he was foot-loose and fancy free. He wandered across New York and into Ohio and Indiana, learned printing in Detroit, and went south to New Orleans. He worked on the Mobile *Register,* then journeyed up through some of the seaboard states, stopping for a spell to run a stage line in North Carolina. He took a job in New York City and later in Washington, where he worked for Duff Green on the *United States Telegraph* and for Gales and Seaton on the *National Intelligencer.* While working on the latter paper he formed a friendship with

[5] *Daily Picayune,* 6 July 1856 (letter from G.W.K.).
[6] *Daily Picayune,* 10 Nov. 1867.

Francis Asbury Lumsden, a North Carolinian nine years his senior.[7]

Kendall's long period of preparation for his real career was almost at an end. In 1835 he and Lumsden were both in New Orleans, working on different newspapers. This was the day of the penny paper: Day's *Sun* had been launched in 1833 and the idea of a cheap press was spreading fast. New Orleans as yet had no such paper. Kendall and Lumsden joined forces, and, on a shoe-string, began publication in January, 1837, of the New Orleans *Picayune,* a saucy little sheet which sold for its namesake, a Spanish coin worth six and one-quarter cents. It was an immediate and enduring success. Its witty cheerful tone, its political independence, its spirit of good will, its emphasis on local news, its refusal to be drawn into controversies with other papers and with individuals, its devotion to the city of New Orleans and the state of Louisiana, attracted a growing list of readers and advertisers.[8]

Four years after the founding of the *Picayune,* there came for Kendall an adventure which was to bring his name and that of his paper into wide renown. Mirabeau Buonaparte Lamar, president of the Republic of Texas, avid for the growth of his country, conceived of an expedition from Texas to Sante Fe to attract for Texas the Santa Fe trade which was doing so much for St. Louis, and to encourage New Mexicans to renounce the sovereignty of Mexico and to adhere to Texas.

Kendall determined to go part of the way with the expedition and then to leave it for a tour of parts of Mexico. The expedition attracted him for a number of reasons. He was always a man of action, a lover of physical activity and ad-

[7] Copeland, *Kendall,* 3–19.

[8] John S. Kendall, "George Wilkins Kendall and the Founding of the New Orleans *Picayune,*" *Louisiana Historical Quarterly,* 11:261–85 (April, 1928). See also Copeland, *Kendall,* 20–32, and Thomas Ewing Dabney, *One Hundred Great Years, the Story of the Times-Picayune from its Founding to 1940* (Baton Rouge, 1944), 1–31.

venture. Then, too, he was a good reporter, and there was promise of something new to write about. Finally there was the strong desire to learn something about the still largely unknown region to the West which he even then foresaw as a part of the United States and a contributor to the prosperity of New Orleans.

The expedition, naturally enough, aroused the fears and hostility of the Mexicans. The up-shot of the matter was that Kendall found himself in prison in Mexico City after a gruelling overland journey of 1,200 miles replete with horrors. Released after a seven months' captivity through the intervention of the United States government, he wrote from memory a long series of articles on the expedition for the *Picayune,* and in 1844 Harper published in two volumes his *Narrative of the Texan Sante Fe Expedition,* a tale of high adventure which sold 40,000 copies in eight years' time. "If there were never such a place as Sante Fe, the narrative would have its charm as a fascinating fiction," said the *Picayune* after Kendall's death.[9]

An enthusiastic supporter of the admission of Texas to the Union, Kendall was in the new state as a reporter when the war with Mexico broke out. Hurrying to the border he began the reporting which perhaps entitles him to rank as the first modern war correspondent. His most notable work was done when he went with the army of Scott from Vera Cruz to Mexico City, attached to the staff of General Worth, sending back to his paper detailed reports of what he saw and heard. Getting the news through was no mean achievement. He hired his own couriers and they raced on their dangerous mission through country alive with bandits. The army itself on occasion made use of "Mr. Kendall's express" to get its despatches through. When Kendall left for New Orleans a few weeks after Scott's entrance into the capital of Mexico, General Worth wrote to him expressing his "high

[9] Copeland, *Kendall,* 42–107; Dabney, *One Hundred Great Years,* 58–64; *Daily Picayune,* 10 Nov. 1867.

and grateful appreciation" of the journalist's services on his staff.[10]

Kendall now determined to publish a book on the war. Before he left Mexico he and Carl Nebel, a French artist, made an agreement to do the work together, Kendall to supply the text for Nebel's illustrations.[11]

In the spring of 1848 he was in France to vacation and to work on his book. But he was a reporter above all else and he arrived in Europe during exciting times. It was more than seven years before he came back to the United States to live (he made several short trips back during that time). The reputation which he had made as a reporter during the war was now enhanced by the distinguished despatches which he sent to his paper from Europe. He witnessed and reported on the Chartist denouement in London,[12] and was in Paris in time to write the story of the bloody June days of 1848.[13] For years his despatches from the French capital gave detailed pictures of French politics and Parisian life. From May, 1848, through December, 1849, the *Picayune* published some 115 letters from him, totalling about 400,000 words.[14] Over the years his letters came from many parts of Europe, from London, Dublin, Hamburg, Berlin, and elsewhere.

Something happened which was not, perhaps, in his plans when he left New Orleans. He fell in love. The girl was just eighteen and he near forty when they met. She was Adeline de Valcourt, whose father had served the first Napoleon and whose mother was of a noble family. She loved him, it may be, for the dangers he had known, and he "loved her that she

---

[10] Copeland, *Kendall*, 140–238; F. Lauriston Bullard, *Famous War Correspondents* (Boston, 1914), 351–74; John S. Kendall, "George Wilkins Kendall and the Founding of the New Orleans *Picayune*," *Louisiana Historical Quarterly*, 11:261–85 (April, 1928).

[11] Copeland, *Kendall*, 229.

[12] *Daily Picayune*, 10 May 1848 (letters from G.W.K.).

[13] *Daily Picayune*, 23 July 1848 (letters from G.W.K.).

[14] Dabney, *One Hundred Great Years*, 82.

did pity them." Before 1849 ended they were married, and in time settled in a Paris suburb. The marriage was kept secret from all but two of Kendall's American friends and relatives for several years, until the death of his Presbyterian mother, who would have been distressed by her son's marriage to a Catholic.[15] Before he returned to the United States to live, Kendall was the father of four children. And he had gone to France to work on a book!

The book got done, too. *The War between the United States and Mexico Illustrated* was an impressive achievement. Twelve illustrations by Nebel covered the war from the battle of Palo Alto to Scott's entrance into Mexico City; the illustrations, as accurate as the collaborators could make them, were lithographed and hand colored in Paris. Kendall's text ran to some 100,000 words. Appleton & Co. published the book, which sold at prices beginning at $36.[16] Well might the author and artist "boldly assert that no country can claim that its battles have been illustrated in a richer, more faithful, or more costly style of lithography."[17]

The years of his European sojourn saw also the beginning of his sheep operations in Texas. Friends were associated with him in his first venture. Although Kendall had a deed for 4,000 acres of land beyond the frontier, which he had acquired in 1845, his flocks first grazed along the Nueces River, above San Patricio. The first purchases were of Mexican ewes from the Rio Grande; to these were soon added six pure blood merino rams from Vermont, including the nine-month-old "Old Poll" which figures so prominently in Kendall's letters, and eighteen pure blood merino ewes from the same state. Kendall made the most of his opportunities in Europe. He bought sheep at the famous farm maintained by the French government at Rambouillet. He hunted in Scotland for a shepherd, finally finding among the Cheviot

---

[15] Copeland, *Kendall,* 248–49.
[16] Copeland, *Kendall,* 252.
[17] Preface.

Hills the young and experienced Joe Tait, who was soon on his way to Texas. From Britain, too, came sheep dogs, and even a couple of Leicester sheep.

The land along the Nueces was good sheep land, but it was not the region to which Kendall hoped to bring his family. The land which he already had to the westward was not to be thought of for the moment as a place to live, because of the danger of Indians. A search began which took him on horseback over a wide area, along the lower Guadalupe and the lower Cibolo and the streams which fed it. Still he found nothing to his liking. Then in the hilly region above New Braunfels, a flourishing little German settlement, he found what he was looking for. "This section," he wrote later, "high, dry, coated with short grass, and without any low or hog-wallow prairie—struck me as not only possessing the advantages I sought in the way of a healthy home, but as being admirably adapted for sheep raising, and here my flock was brought about the commencement of 1853." [18]

These early ranching years, with Kendall spending most of his time abroad, were years of tribulation. A severe storm killed many of the unsheltered sheep. Then the flocks were attacked with liver rot, and again many sheep perished. He was interested to note, however, that the disease attacked mostly those sheep which had been moved from the lower country; those which he had bought in the New Braunfels region were largely untouched. He decided to begin grazing sheep on his lands thirty miles to the west, near Boerne. A prairie fire swept over the new range and a careless shepherd let his charges get caught in it. Between 300 and 400 sheep died as a result of burns. That same winter the lambs started coming too early, before new grass had started. During the following fall and early winter many lambs died. Then came a hard winter. Cold, wet northers blew almost constantly from January through March, 1856; fires the preceding fall

---

[18] George Wilkins Kendall, "Sheep Raising in Texas," *Texas Almanac for 1858,* 134–36; Copeland, *Kendall,* passim.

had destroyed prairie grass, and there was little fodder for the cold and hungry animals; again the losses were heavy. It was at this point that Kendall and his family came to Texas to live.[19]

Despite the troubles of the preceding years the venture had survived, and Kendall had learned a good deal about sheep and their care. Better watching, provision of emergency fodder, care in safeguarding against too early lambs, provision of special shelters for the weak and ailing—these things a determined man could see to, and Kendall was determined. As he himself put it, few would "watch as close and work as hard."[20]

Years of hard work rewarded by gratifying success followed. He had soon cut his losses down to less than one per cent—this at a time when flock masters counted themselves successful with losses of fifteen and twenty per cent. The eyes of Texas, and of other states and of foreign countries also, were upon him. Said the *Texas Almanac for 1859:* "If Mr. K can go on for another year, with the same extraordinary success which has attended him the two just past, he will have incontestably proved the fact that no better sheep range exists in the wide world than can be found in the mountains of Comal, Blanco, Hays, Gillespie, Kerr and Bexar counties, and, perhaps, even wider limits of that section may be taken in."[21]

While becoming, in the words of Edward Wentworth, "the greatest sheepman Texas ever claimed,"[22] Kendall also became one of its greatest publicists. His enthusiasm for Texas was unbounded—exaggerated, a non-Texan might some-

---

[19] Kendall, "Sheep Raising in Texas," *Texas Almanac for 1858*, 134–36.

[20] See below, letter of 19 March 1860.

[21] P. 196. For Kendall's own reports see *Texas Almanac for 1858*, 134–36; *Texas Almanac for 1859*, 126–28; *Texas Almanac for 1860*, 157–59; *Texas Almanac for 1861*, 166–70.

[22] Edward Wentworth, *America's Sheep Trails* (Ames, Iowa, 1948), 382.

times think. His writings for the *Picayune* and for the *Texas Almanac*, and his private correspondence helped to focus on him the attention of multitudes interested in settling in Texas. When the Boston *Post* published a letter which he had written to a friend, Kendall himself received "at least three hundred letters." The last mails, he reported six months later, had "brought letters from California and the Sandwich Islands." [23] He answered them all with large expenditures of time and money. Later he had the *Picayune* print copies of a letter on Texas which he wrote, and for a time he sent one of these in response to an inquiry.[24] The letters, however, were only half the story. Untold numbers of people searching for opportunities in Texas passed through his most hospitable door. The demands on him had not ceased by 1860 as the following letters show.

He had a name for those who belittled Texas; they were "croakers" and his contempt for them was great. He well knew that men might fail in Texas, but he believed that the fault was likely to lie with the men. In 1859 he wrote: "I have never been in a country where so little work was done as in Texas; I have never been in a country where a man would gain as much by work as here." [25] As for him, he proposed to stay until the Guadalupe started to run dry, and then he would leave, his "station being at the last or upper water hole, bringing up the tail of the mourners." [26]

Although he believed in slavery as an institution, defended the doctrine of states rights, and hated abolitionists and Black Republicans, Kendall was at the same time no extremist. In 1860 and early 1861 he felt that the headlong flight from the Union of South Carolina and her sisters of the Deep South was premature, that Virginia and Kentucky, on the

---

[23] *Daily Picayune,* 27 July 1858 (letter from G.W.K.).

[24] Copeland, *Kendall,* 286–87.

[25] *Daily Picayune,* 2 Sept. 1859 (letter from G.W.K.).

[26] *Daily Picayune,* 28 June 1857 (letter from G.W.K.).

very border of Black Republicanism, should be the leaders, their provocation being greater. Nevertheless, when Texas made her decision he was willing to accept it. "I reasoned," he wrote, "that we were all at sea in the same boat, with breakers ahead, and that however much some of us might differ from our pilots, it was the duty of all to bend lustily to the oars, and endeavor to gain a common harbor of security and safety." [27]

By May, 1861, after Sumter and Lincoln's call for volunteers, he was in a different mood. "The deceit and double-dealing of Abraham Lincoln—duplicity which would disgrace the Mexican Santa Anna, be a match for Satan himself, and shame even a Camanche—his recent course, I say, virtually declaring war against the South, has gone far towards changing my mind, and in the end perhaps the course pursued by the Confederate States will be the wiser." He now suggested a vigorous program for carrying on the war: direct taxation and a pay-as-you-go policy, long term or duration enlistments for the army, enlistment of friendly Indians.[28]

He spent the war years on his ranch at Post Oak Spring faced with endless problems and annoyances. He took an oath of allegiance to the Confederacy only when he had to in order to market some sheep in Mexico.[29] Not once did he get to his beloved New Orleans.

In 1861 Texas had no woolen mills. At the beginning of the war Texas wool was bought by the Confederate government and sent east, to Georgia, Virginia, and other states for manufacture into the blankets and clothing so badly needed by the armies. It cost, Kendall estimated, almost 30 cents to get a pound of wool from Texas to the eastern mills, and as much to bring back the finished goods. Results

---

[27] *Daily Picayune*, 19 May 1861 (letter from G.W.K.).
[28] Ibid.
[29] Copeland, *Kendall,* 299–300.

were not always good either; machinery in the east faced wool shortages while trans-Mississippi soldiers were short of blankets and clothes.

Kendall wrote a letter to the Quartermaster-General of the Confederacy proposing the establishment of a woolen mill in Texas. At New Braunfels, he urged, there was water power aplenty, a vacant floor in a mill building, and access to a labor supply.[30]

The Quartermaster-General never responded to the letter. The marketing of Texas wool became a problem. Once after the war started, Kendall sent a large shipment to New Orleans. The occupation of that city by Union forces during the early part of 1862 made it impossible to repeat this performance. Thereafter his wool went to Mexico, to Shreveport, or remained in storage.[31] The fall of Vicksburg completed the isolation of Texas. Wool in Texas became a drug on the market. Government wool remained in storage, wool in private hands was not always sure of a shelter. Kendall noted toward the end of 1864 that whereas he had sold his clip in 1860 for 28¼ cents a pound, no one was then offering him a third of that.[32] Texas wool growers, he thought, suffered more as the result of the war than other classes.[33]

The end of 1861 found his flocks in good condition. There had been difficulties with wolves and wild cattle, and with his shepherds. The unwillingness of his Mexican shepherds to take paper money led to trouble and the shepherds left.[34] Finding new shepherds was no easy task.

Then disaster came. The Comanches had long been a dan-

---

[30] Kendall, "Sheep Raising in Texas," *Texas Almanac for 1865*, 39–42.

[31] Copeland, *Kendall*, 296, 300, 301.

[32] *Texas Almanac for 1865*, 40.

[33] Kendall, "Sheep Raising in Texas," *Texas Almanac for 1867*, 217–19.

[34] Copeland, *Kendall*, 297.

ger on the Texas frontier, but thus far Kendall had escaped their depredations. In the first month of the new year his luck ran out. With "one murderous dash" Comanche raiders killed three of his shepherds. His description of the search for the shepherds and the finding of one of the bodies brings us close to the horrible reality. He wrote while the nightmare was still fresh:

After dark—and it was pitchy dark—Putnam and myself, well armed, went over to the point where he had left Fechler on foot; we shouted again and again, hoping that he might be hidden in the bushes, or perhaps wounded and unable to move; but no response came through the darkness. We next went below the wether pen, in the direction which Baptiste must come, and shouted our loudest. No sound, not even the echo of our own voices, came back through the dark and damp gloom. We next struck over to the pen where Schlosser's flock was folded; neither sheep nor shepherd were there; nor was any response given to our repeated calls for the unfortunate man; all was silent save the screeching and hooting of owls perched in the neighboring trees.

Coming back to my house, about 9 o'clock, I sent a man over to Boerne, on a fast horse, with a note to our Senator, Mr. Reed, asking him to collect all the men he could for work in the morning. At the same time I despatched Mr. Putnam in the direction of the Guadalupe, to arouse the neighbors in that quarter. In such a pitchy black night nothing more could be done; but dark and gloomy as was the night, my own forebodings were even more dark and gloomy. I tied my horses close in to the house, and spent the night in watchful anxiety for the coming day.

At an early hour on Saturday morning—the weather still being dark, damp and foggy—Putnam came in with my old head shepherd, Tait. Going to the spot where the former had first seen the Indians, the body of the poor boy Fechler was found, stripped naked, pierced by some seventeen arrows, and his head doubled under him and resting against a tree. His scalp had not been taken, nor his person mutilated save by his many wounds; yet the dreadful spectacle called up mingled feelings of deep pity for the unfortunate lad, boiling indignation against his brutal murderers, and deep-rooted disgust at the majority of our rulers who have never bestowed a second thought upon frontier protection.[35]

It is understandable that Kendall, in the face of provocation such as this, was, in his view of Indians, wholly and simply a man of the frontier, favoring a war of extermination, scornful of "mawkish philanthropists" [36] who tried to make a case for the Indians, furious with politicians who were oblivious to the cries of the settlers for stronger defenses.

The Indians [he wrote] must be worried and hunted down, killed to the last man, or driven so far beyond our borders that we shall hear of them no more. They should be sought in their hunting grounds, when killing their stock of buffalo meat—should be allowed no rest nor respite—should be granted no truce nor treaty. The cruel war they are waging against us—and cowardly as well as cruel, for they attack none but the weak and defenseless—will last so long as an Indian is left. This is a fact we must look in the face, and must act upon it, or else abandon the frontiers sooner or later.[37]

The Indian foray made life harder for all. New shepherds were hard to find "while the blood of the old was still fresh on the ground." When found, they were fearful of taking their flocks out far on the range, and the flocks began to suffer as a result. Farming and gardening operations were for the time abandoned. The children were scarcely allowed to play around the yard.[38] And although men of the section could ill be spared from home, Kendall and his neighbors organized and carried on scouting expeditions.

The body of the third shepherd had not yet been found when a fresh danger arose. Fires swept the prairies, endangering the flocks, two of Kendall's being saved only with

[35] *Daily Picayune*, 5 Feb. 1862 (letter of George Wilkins Kendall to the San Antonio *Herald*).

[36] *Daily Picayune*, 28 Nov. 1858 (letter from G.W.K.).

[37] *Daily Picayune*, 12 Feb. 1862 (letter of George Wilkins Kendall to San Antonio *Herald*).

[38] Kendall, "Sheep Ranching in Texas," *Texas Almanac for 1863*, 40–41.

great effort. Miles to his east and west the prairies stretched, "black and desolate." [39]

A winter filled with disaster and forebodings was followed by a cold and bitter spring. Losses were much greater than usual. The summer was hot and dry, and the winter set in early. But despite the wretched months behind, Kendall's flocks were in good shape in November, 1862, and he was optimistic for the future, provided he could secure good undershepherds; in his head shepherd, John McKenzie, he had "a faithful and trusty Scotchman on the watch." He was not prepared to give up. "But come what may," he trumpeted, "my faith in Texas as one of the best wool growing regions on the face of the earth is firm and abiding, and to this faith I shall cling even beyond the bitter end." [40]

At the end of another year, Kendall felt that he could not reasonably complain "spite of drawbacks and annoyances almost innumerable." There was the eternal problem of getting help. When he could not find another shepherd, he had to make two lamb flocks out of three, making about 1,700 for each flock, an excessive number for the best results. At shearing time there had come "countless myriads of flies . . . followed by armies of worms." The danger of scab was constant; it was everywhere around him.[41]

The year 1864 was not cheerful. Kendall lost a good many lambs during the cold spring when they were being dropped. His spring clip went unsold. And before the year's end a stray sheep from a neighbor's flock got into one of his flocks, and soon two of his flocks were infested with scab. At first he hoped to avoid dipping until spring, after shearing was over. Then he would go to work. "As the overanxious candidate on the morning of the election called upon his friends

[39] *Daily Picayune,* 12 Feb. 1862 (Kendall's letter to San Antonio *Herald*).

[40] *Texas Almanac for 1863,* 41.

[41] George Wilkins Kendall, "Sheep Raising in Texas," *Texas Almanac for 1864,* 37–39.

to vote twice, to vote often, to keep on voting until the polls were closed, in order to make sure of his return, so will I dip twice, dip often, keep on dipping until I can say that there is not a sign or vestige of the scab on my place." [42]

It was soon clear to him that he could not wait until shearing time to dip his sheep. Nursing them through the winter through attention to individual sheep would be too staggering a task. Weeks of preparation followed his decision not to delay. He bought sulphur, gathered rocks and wood (which were burned to obtain lime and lye), borrowed utensils. Then came the dipping, which went on for more than two weeks before his 5,000 had been treated; many died from the treatment.[43]

More troubles lay ahead. A storm killed nearly a hundred lambs, and hard on its heels came another in which a flock was lost on the range and partly destroyed. Its shepherd was frozen to death. With spring came the need of renewed dipping, an operation which this time was carried on in a vat which Kendall had had constructed for the purpose. Help was still scarce in the face of Indian threats and reports of the operations of "bushwhackers." Kendall himself spent the nights with one of his flocks, and his older boy, now thirteen, went out with another.[44]

But there was a brighter side to things now. The war was over. Wool would again become a saleable commodity. Best of all, at long last, it was possible to make again a trip to New Orleans. Georgina, now fifteen, was to be placed in a school there.

The return to the city of his heart was both joyous and sad. New Orleans was gay, and the Kendalls attended the theatre almost nightly. Among many of his old friends, however, he found little disposition to be gay; the war had brought sadness and failure to them.

Back at home, he began to plan a trip to France to see

---

[42] *Texas Almanac for 1865*, 40.

[43] *Texas Almanac for 1865*, 42; Copeland, *Kendall*, 302–3.

[44] Copeland, *Kendall*, 303–5.

Caroline ("Liline"), his deaf daughter, now twelve years old, who had never seen America. When a Mexican shepherd was killed in self-defense by another shepherd, and all the other Mexicans quit, Kendall had to postpone his departure until the situation could be straightened out. In June, 1866, he and his wife started out, and in New Orleans were joined by Georgina. They made stops at Louisville and Niagara Falls, then went on to Cortland to visit Henry Randall and his family. In New York City they saw Fletcher Westray and other friends, and then sailed for Europe. They were soon back in New York with Liline. Kendall and Georgina visited relatives in New England; then the family, after a short stay in New York, headed for Washington and New Orleans.[45]

Henry Randall has left us a pleasant picture of Kendall at Cortland. He wrote more than a year after the visit:

We seem to see and hear him now at the table, in the drawing-room, on the ride, and in the walk. How often we thought of Charles Lamb! His manner was not that of a professional wit. He was not "reminded of a good story" by everything which was said and done. He did not seize on every available opportunity. The flash came suddenly and as if inadvertently. The "good thing" was often uttered *sotto voce* only to the person in the next chair, and without any of the preparatory signals by which old stagers in such matters generally bespeak the attention of the audience. He did not seem to be witty on purpose—but only because he could not help it.[46]

The Kendalls spent the winter of 1866–67 in New Orleans. Kendall spent long hours at the *Picayune* Office, sometimes not leaving until after midnight.[47] He found that over the years he had drawn out more than his share from the partnership.[48] The absence of one partner and the frequent ill-

---

[45] Ibid., 305–10.

[46] *Moore's Rural New Yorker*, 2 Nov. 1867. This was reprinted in the *Texas Almanac for 1868*, 178–79.

[47] Adeline de Valcourt Kendall to Henry S. Randall, 29 Jan. 1867, New-York Historical Society.

[48] Copeland, *Kendall*, 315.

ness of another made more work for him. "Mr. Kendall could be seen," a friend wrote later, "at his desk or in the nearest corner where he could find a place for his papers, busily absorbed in working out the great problems with which his brain teemed, of how to invigorate the industry, and lift up the fortune of Texas and New Orleans." [49]

He was eager to get to his ranch. "Randall," he wrote in January, 1867, "I am going down the Western slope of life— cannot carry the weights, run the races, nor pack the loads I once could. With Texas air, Texas exercise and Texas habits generally—active, out-door and bracing—and I can go on for years barring accidents and ordinary ills; but here, confined to the office, I am daily admonished that it is necessary to lock or put the brakes going down hill: in other words, to use all caution and prudence in eating and drinking." [50] Mrs. Kendall was well aware of his impatience to be off. "His love and admiration for that State increases, I believe, every day," she wrote to Randall. She, poor woman, did not share his impatience to return to the privations of ranch life; but good wife that she was, she was ready to go back whenever he gave the word. [51]

Although Kendall might think so, it was not "in the cards" for him to win again. The summer and early fall at the ranch were hard for him, but to Mrs. Kendall at least his health never seemed better than ten days before his death. The two of them made a trip to San Antonio to see the boys. They were scarcely home again before illness came which ended in his death in a week's time. A doctor who was visiting a neighbor eight miles away tried to help; a second doctor, from San Antonio, arrived with the boys too late. "It is terrible to live in the woods, away from Doctors, in case of sickness," wrote Mrs. Kendall. [52]

[49] *Daily Picayune,* 10 Nov. 1867.

[50] Kendall to Randall, 30 Jan. 1867, New-York Historical Society.

[51] 29 Jan. 1867, New-York Historical Society.

[52] Mrs. Kendall to Holbrook, 27 Oct. 1867, printed copy, New-York Historical Society.

Henry Randall wrote for *Moore's Rural New Yorker* an obituary of Kendall which was reprinted in the *Texas Almanac for 1868*. Of his departed friend he said: "He loved Texas with an absolute devotion. He never was tired of writing or speaking in its praise. He loved its vast expanses of solitude, its majestic plains, its noble rivers, the green hills of the county named after him, and its masculine energetic population. Texas will deeply miss and mourn him. Perhaps she had no citizen which she could have so ill afforded to spare. She certainly has none who can entirely fill his particular place." [53]

"A sheep ranche cannot be neglected," wrote Mrs. Kendall a few days after her husband's death. With the aid of her brother-in-law (Henriette's husband), Georgina and the boys, she undertook to carry on. [54] Six years after Kendall's death she married Benjamin F. Dane, one of the men who had looked after Kendall's sheep while he was away during 1866–67. She survived until 1924. [55] A cultured and sensitive woman, she had made the transition from Paris to a ranch in Texas with bravery and devotion.

Henry Stephens Randall, Kendall's correspondent, was no less distinguished. About two years younger than his Texas friend, he was by 1860 a leading authority on sheep husbandry in the United States. He was the author of a monumental life of Thomas Jefferson which, even today, remains the most extensive of the Jefferson biographies. He was a prominent Democrat, active in local, state and national politics. A strong friend of public education in his own state, his work had brought him offers of positions of superintendent of public instruction from other states.

Randall had the advantages of membership in a well-to-do, influential family. Born in Madison County, New York, in 1811, he had moved with his parents to Cortland two years

---

[53] *Moore's Rural New Yorker*, 2 Nov. 1867.

[54] Mrs. Kendall to Holbrook, 27 Oct. 1867, printed copy, New-York Historical Society.

[55] Copeland, *Kendall*, 319–20.

later. His father and uncle soon became men of mark in the young community. By the time Henry graduated from Union College at nineteen, his father owned two stores, a hotel, a fine brick residence and a good deal of land. The elder Randall, who lived elegantly and ostentatiously, became the subject of village legend.[56]

After his graduation from college young Henry read law, was admitted to the bar, and pursued a number of interests. He married in 1834, and his father built for him a substantial frame house next to his own. Acquainted with sheep from boyhood, he now became engrossed in learning as much about them as he could. He was active in the New York State Agricultural Society and helped organize an agricultural society for his county. In 1838, as chairman of a committee, he presented at a meeting of the state society a paper on the sheep of the United States.[57] Looking back years later to the organization of the Cortland County society, he described himself as having then been, "a young lawyer and editor, brimful of zeal, with a little book knowledge, and of course a much better understanding of farming, in his own opinion, than he possessed twenty years afterwards." [58] He bought and edited a Democratic paper. In 1835 he was the youngest regular delegate to the Democratic National Convention. He became a school visitor for his county.

During the next decade he became more widely known. He was corresponding secretary of the state Agricultural Society, an office which brought him into contact with many agriculturalists far beyond the boundaries of New York. He worked vigorously for the success of the first annual fair sponsored by the Society. In 1844 he was awarded at the fair first premiums for the best merino buck and ewes. He wrote

[56] Bertha Eveleth Blodgett, *Stories of Cortland County* (Cortland, 1952). There has been no full length biography of Randall. The *Dictionary of American Biography*, XV (1935), has a sketch.

[57] For the paper see the *Cultivator*, V:24–26 (March, 1838).

[58] *Transactions of the New York State Agricultural Society* for 1862 (Albany, 1863), 343.

letters on sheep raising for the Virginia *Valley Farmer* in 1845, and in the same year wrote for Secretary of the Treasury Walker a report on New York agriculture.[59] For *The Farmers' Library and Monthly Journal of Agriculture* he wrote a series of articles which in 1848 were published as a book of about 300 pages—*Sheep Husbandry in the South*, destined to be reprinted several times.[60] He began, also, to send "colonies of Merinos" to many parts of the South.[61] For several years he was county superintendent of common schools; in 1843 he wrote a sixty page article on "Common School Libraries." [62] In 1848 he was a Free Soiler, writing the Buffalo declaration of that party.[63]

The 1850's found him busier than ever. He served an elective term as Secretary of State in New York, a position which gave him also the state superintendency of public instruction, which was his real interest. The *Report of the Commissioner of Patents* for 1850 carried an article by him on "Sheep Husbandry and Wool-Growing in the United States." [64] His great work during this decade, however, was his life of "St. Thomas of Monticello," which he labored on from 1853 to 1858.[65] In writing the three volumes he had the full co-operation of Jefferson's family, who sent him man-

---

[59] *Reports of the Secretary of the Treasury of the United States,* V (Washington, 1851), 309–22.

[60] In the later printings the title was reduced to *Sheep Husbandry.*

[61] Henry S. Randall, "Sheep Husbandry and Wool-Growing in the United States," in *Report of the Commissioner of Patents, for the Year 1850,* Part II, *Agriculture* (Washington, 1851), 140.

[62] Samuel S. Randall, ed., *Mental and Moral Culture, and Popular Education* (New York and Boston, 1844), 176–236.

[63] DeAlva Stanwood Alexander, *A Political History of the State of New York* (New York, 1906–1909), II, 324.

[64] *Report of the Commissioner of Patents, for the Year 1850,* Part II, *Agriculture,* 129–44.

[65] Henry S. Randall, *The Life of Thomas Jefferson* (3 vols., New York, 1858). The reference to Jefferson as "St. Thomas" appears in a letter written by Randall to Albert W. Bishop, 11 July 1859, in the possession of Mrs. Arthur Dunn, Cortland, New York.

uscripts and even allowed him to keep some of them. It is easy to understand why Randall, a man of large intellect and varied interests, a Democrat, a friend of public education, a farmer and a patriot, should have found his hero in the great Virginian.

For Randall the period from 1861 to 1867 was one of tremendous activity. Much of this activity was in behalf of the sheep industry.

While Texas wool growers watched their markets shrink and their prices decline, Northern growers as the result of war found a vastly expanded market and improved prices. "If cotton was king, it is king no more, and wool is now the cry," remarked one observer. The downward trend in sheep raising in the older states which had set in before the war was for the moment reversed. Prices of blooded stock went sky-high. The heydey of the Vermont breeders had arrived. Sales of rams at $5,000 and ewes at $3,000 were reported.[66] George Campbell took two first premiums and one second premium with twelve merinos which he took to the great international agricultural exposition at Hamburg in 1863, and then sold the sheep to a Silesian nobleman for $5,000.[67] Edwin Hammond, it was said, could have sold his ram, Gold Drop, for $10,000.[68]

Randall continued to engage in the sheep business. Although his southern market had of course disappeared, the increased activity in the North had doubtless more than compensated. In 1864 a visitor to Vermont found that Randall had just left for home with sheep which he had bought of Hammond. The same writer was soon afterwards at Cort-

---

[66] Henry S. Randall, "Recent Improvements in American Sheep, Etc.," *Texas Almanac for 1868,* 162.

[67] *Report of the Commissioner of Agriculture for the Year 1863* (Washington, 1863), 28–29.

[68] *Prairie Farmer,* 26 Aug. 1865, 149. Gold Drop, who died prematurely in 1865, was said to have been valued by Hammond at $25,000.

land, and reported that Randall had a flock of some 800 sheep, "all thoroughbred and of fine quality." [69]

His pen was seldom idle during the period. In 1862 he read before the New York State Agricultural Society an elaborate essay which was soon published as a book, *Fine Wool Sheep Husbandry*.[70] An article by him entitled "Sheep," appeared in the *Report of the Commissioner of Agriculture* for 1863.[71] In 1863 also appeared his best seller, *The Practical Shepherd*, a guide to sheep raising which went through many printings. "It's the *Sheep Bible*," one of its publishers declared,[72] and Randall began to be referred to as *"the* authority" on sheep. *The Practical Shepherd* was the first of his writings on sheep which had ever brought him any money; until then he had always refused to accept pay for his work. In 1864 he became the paid conductor of a weekly column, "Sheep Husbandry," in *Moore's Rural New Yorker*, an excellent farm paper published in Rochester, New York, which enjoyed a good circulation. In discussing his writings on sheep Randall estimated in 1864 that his letters, addresses and so on which had been printed, equaled in quantity the material in his three books, and that the material in his unprinted letters was greater than that of either of the classes of published matter. "It will therefore be conceded we think," he wrote, "that we have ridden our hobby *hard*, whether we sit him well or not!" [73]

His hobby took him during this period in a new direction. He became the leading spirit in organizing the wool growers and battling for greater tariff protection. He wrote much on

[69] *Ohio Farmer*, 17 Sept. 1864.

[70] The essay was published in the *Transactions of the New York State Agricultural Society* for 1861 (Albany, 1862), 663–784. As a book it was first published in Albany in 1862.

[71] Pp. 229–47.

[72] D.D.T. M[oore] to "My Dear Doctor," [Randall], 18 May 1864, Randall Papers. *The Practical Shepherd* was published in 1863 in Rochester by Moore, and in Philadelphia by Lippincott.

[73] *Moore's Rural New Yorker*, 6 Aug. 1864.

the subject. He urged that Congressmen be inundated with petitions. He promoted and became president of the New York Sheep Breeders' and Wool Growers' Association. He met with prominent manufacturers in a conference which was the prelude to an alliance between the two interests. He presided at the famous Syracuse Convention which cemented the alliance. He became the president of the National Wool Growers' Association, organized on the eve of the meeting at Syracuse. He headed the committee of growers which met with the manufacturers to reach an understanding in respect to a new wool tariff. He spent weeks in Washington while the details of a wool schedule were being worked out. The objective for which he had labored was reached when Congress passed the Wool and Woolens Act of 1867, a measure of great national significance. As a result of the work of Randall and those who fought with him, the wool and woolens industry was to be "the very citadel of protection" for many years to come.[74]

Randall lived until 1876. His work, however, had been largely accomplished before the death of Kendall. There were no more books. He served a term in the New York legislature—a personal triumph since he was a Democrat living in a Republican district. He was chairman of the legislative committee on public education. He served as a trustee of the new normal school at Cortland. He continued to head the National Wool Growers' Association and the New York association as well. He presided at the second Syracuse convention of wool growers and wool manufacturers on the eve of the tariff revision of 1872.

His old friend, D. D. T. Moore, of *Moore's Rural New Yorker,* paid him at the end a high tribute: "Indeed, no man

---

[74] For a detailed account of Randall's efforts in behalf of protection for wool growers, see Harry Brown, "The Fleece and the Loom: Wool Growers and Wool Manufacturers during the Civil War Decade," *Business History Review,* XXIX: 1–27 (March, 1955).

GEORGE WILKINS KENDALL

HENRY STEPHENS RANDALL

in America, and probably none in the world, has done so much to enhance the interests of Sheep Husbandry, as he whose decease is now most regretfully chronicled." [75]

Randall was an intense man, with high white forehead and brilliant eyes, with an enormous capacity for work. "I have seen few men," he wrote in 1856, "who entered an investigation with such zeal as myself." [76] And again: "When I work, I work so fiercely that talking & letter writing is only a rest— an amusement—an unbending to me. I rarely write a letter, at home, except after a hard days work of ten steady hours at least." [77]

As a letter writer he must rank among the most prolific of his day. He wrote multitudes of letters on a variety of subjects. It was nothing for him, he said, to write half a dozen letters to establish a date. He wrote to scholars, journalists, politicians, manufacturers, farmers. He wrote with amazing speed, two of the swiftest clerks in the State Department at Albany being unable to copy his letters as fast as he wrote them—and he sometimes wrote a hundred in a day.[78] He wrote vigorously, bluntly and emphatically; it was a rare letter which did not have underscorings.

Randall and Kendall had more in common than an interest in sheep. They were a good deal alike. Both were men of intellect, with wide veins of down-to-earth practicality. Both were high-strung men who lived to the hilt, fervently and with strong convictions. Both were persons of warmth, with a great capacity for friendship, and the ability to enjoy themselves in a group. Both drove themselves beyond the usual limits of human beings. Both were men of humor, not

---

[75] *Moore's Rural New Yorker*, 26 Aug. 1876.

[76] Frank J. Klingberg and Frank W. Klingberg, eds., *The Correspondence between Henry Stephens Randall and Hugh Blair Grigsby, 1856–1861* (Berkeley and Los Angeles, 1952), 57.

[77] Ibid., 57.

[78] Ibid., 58.

a subtle and precious sort, but the kind that comes with high spirits and a great delight in life. Strong, blunt, honest, successful men, both of them, there was between them a mutual attraction and respect.

Kendall reveals in his letters the pleasure which he took in Randall's letters to him. In his obituary of his Texas friend, Randall had this to say: "As an epistolary writer we never knew his equal. He was our close correspondent for years, except during the war, and the grace, wit and hearty mirth of his letters—their constant novelties of thought and expression—their uniform zest—their diverting illustrations—were all inimitable." [79]

Sheep husbandry was in 1860 an important part of American agriculture, relatively a more important part than it is today.[80] At that time some 24,000,000 sheep fed on the grasses of the United States and yielded perhaps 80,000,000 pounds of wool. Although the industry was found in every state, six states had about half the total number of sheep. Ohio led with about 3,500,000 grazing her rolling lands; New York, Pennsylvania, Michigan, California and Virginia followed in the order named. Vermont with about 750,000 was the source of blooded stock for many a flock in other states.

Since early colonial days sheep had been part of the American scene. The English, the Spanish and the Dutch all introduced them into their American colonies, where they increased in numbers over the years and made a substantial

---

[79] *Moore's Rural New Yorker,* 2 Nov. 1867.

[80] Sources of information concerning the sheep industry during this period include Brown, "The Fleece and the Loom: Wool Growers and Wool Manufacturers during the Civil War Decade," *Business History Review,* XXIX:1–27 (March, 1955); Wentworth, *America's Sheep Trails;* Chester Whitney Wright, *Wool Growing and the Tariff: A Study in the Economic History of the United States* (Boston, 1910); *Eighth Census of the United States, 1860, Agriculture* (Washington, 1865).

contribution to life in the new world. Not until the beginning of the nineteenth century, however, did the modern history of sheep husbandry begin in the United States, when thousands of pure blood merinos were imported from Spain. For a time the Spanish merino craze was unabated, but at length a violent reaction set in. Later came a short-lived enthusiasm for fine-wooled Saxon merinos. Other merino breeds and English mutton-type sheep were also introduced. But the Spanish merino persisted, and in time, through the careful breeding of such men as Edwin Hammond, developed into the so-called American merino. It is safe to say that in 1860 the majority of sheep in the United States had an infusion of Spanish merino blood. The wool produced was largely of a medium quality suited to the needs of the domestic cloth manufacture, whose output was chiefly designed for popular consumption. Some coarse wool, used in blankets, and some extra fine wool was also grown. The deficiency was in very coarse wool and in long combing wool, the former the raw material of the carpet industry and the latter of the developing worsted industry.

The market for the clip was wholly domestic, and in the domestic market the American grower had to compete with foreign growers. During the year ending 30 June 1860, about 26,000,000 pounds of wool were imported, a considerable part of this mestiza wools from Argentina (the product of a cross of the merino with the "native" stock), which were highly competitive. Furthermore, there was the competition of foreign wool imported in the form of manufactured goods. During the year named, manufactures of wool worth more than $43,000,000 were imported. Altogether the American wool grower supplied no more than from one-third to two-fifths of the wool consumed in the United States. Under the tariff act of 1857 most raw wools were imported free of duty, and this was an increasingly sore point with people like Henry S. Randall.

Uniformity did not characterize the industry. In Vermont

a farmer's flock might comprise two or three dozen sheep in a rock-strewn pasture not far from the barn; there shelter and an accumulated store of fodder were necessary during the long cold winters. In California and Texas holdings might run to thousands of sheep, always under the watchful eyes of herders; there rude shelters from the storm and a small store of emergency rations would suffice, or perhaps there would be no shelter or fodder other than what Nature herself might provide at the moment of need. In between these extremes were flocks of varying sizes, the vast majority of them contributing only a portion of the livelihood of their owners. Some sheepmen bred blooded stock for sale. Others raised sheep primarily for what the wool would bring in the market. Still others were becoming interested in mutton sheep, and with them the wool was secondary.

A change in the industry was in progress. A decline was in evidence in most of the older, settled parts of the country. In the newer regions where lands were plentiful and cheap the trend was upward. While the number of sheep fell off in New England and other long settled regions between 1850 and 1860, it increased several fold in California and Texas. For states such as Massachusetts, New York, Pennsylvania and Ohio, where land values were rising, dogs increasing and winters rigorous, the dairy industry was becoming a more profitable activity. In the older southern states wolves, slavery, parasites and the emphasis on cotton contributed to the decline. These regions had all once been frontiers, but that day had passed and new frontiers were eagerly grasping for an industry which they seemed particularly qualified to develop.

Most of the woolen mills of the country were to be found in New England, New York, New Jersey and Pennsylvania. Wool brokers and wool merchants flourished in Boston, the great wool marketing center, and in other cities. Manufacturers bought through them or direct from the producing areas. Slipshod and even dishonest practices attended the preparation of wool for market: the dirtiest and coarsest parts of fleeces were often mixed with the cleanest and best; an excessive amount of twine was sometimes used to tie the fleeces, since twine was cheaper than wool; the twine was frequently of a type which resulted in imperfections in cloth when it had by chance become mixed with the wool in the factory; stones were even introduced on occasion to make a lot of wool weigh more. The growers had their complaints also: to them it appeared that the buying practices of manufacturers were designed to depress the price of wool.

This was the industry of which both George Wilkins Kendall and Henry S. Randall were a part. Kendall was a man of a new frontier, a creator, a builder of a new economy in a new commonwealth. Randall was a man of an older region, who, through his advice, based on long experience and study, and his shipments of sheep, was making no small contribution to the development of far-away Texas, a land he never saw.

THE LETTERS OF 1860

Rancho near New Braunfels
Jan. 1, 1860.

A Happy New Year to you, my dear Mr. Randall, and to all your household, and many happy returns. The new year comes in upon us here bright and beautiful, with a clear sky and bracing wind: it finds us all in good health, and all, I trust, duly thankful for the many blessings we have enjoyed the past year. If 1860 will do as much for us, we shall all be satisfied.

I came down from Post Oak Spring yesterday: left all my flocks in the same fine condition—all healthy and thriving. Although a Norther [1] visited us during the three days which wound up the old year, with occasional spits of sleet, my sheep were not affected: total loss during the last month, *two*—both late lambs, and stunted for want of milk. I have been lucky.

Meanwhile the loss of sheep in Texas this season, during the storms, has been awful, mostly among the Mexican and Missouri new comers. One gentleman in San Antonio told me that a friend of his, just in from the Northeastern coun-

---

[1] Northers are winds or storms coming from the north to which parts of Texas are subject during the fall, winter and spring. They are usually accompanied by intense dryness, although "wet" northers do occur. Temperature falls sharply during such a storm. During the winter of 1859–1860, twenty-eight northers occurred in Kendall's region, the first at the end of September and the last toward the end of April. Their average duration was four days.

Frederick Law Olmsted in *A Journey Through Texas* (New York, 1857), 99, thus describes the onset of his first norther: "The air had been perfectly calm; but as we arrived near the next summit there was suddenly a puff of wind from the westward, bringing with it the scent of burning hay; and in less than thirty seconds, another puff, chill as if the door of a vault had been opened at our side; a minute more, it was a keen but not severe cold northerly wind. In five minutes we had all got our overcoats on, and were bending against it in our saddles. The change in temperature was not very great (12° in 12 minutes) but was singularly rapid; in fact instantaneous—from rather uncomfortably warm to rather uncomfortably cool."

ties, saw no less than 700 dead sheep stacked up in a pile, which they were about to burn, all lost in a single night! Another man lost 200 Merinoes in a night, all frozen: a poor, hard-driven sheep, cannot stand such Northers as we have had.

On looking into the open pens where my two ewe flocks are kept, yesterday morning, a cold sleet cutting my face and ears, I noticed that every sheep was quietly and complacently chewing her cud, and apparently contented with her lot in life. All are with lamb, nearly 2000 of them, and if I get through the winter without great loss, and the ewes in fair condition, I shall be able to show you a pleasant sight next May. In my day I have witnessed the entrechats and pirouttes [sic] of the best ballet troupes in Paris, London, Berlin, New York, Havana, Mexico, and other cities: but the skipping and playing of a couple of thousand lambs of an evening, all fat, full of milk and full of spirits, is a sight more pleasing to me—a heap more pleasing, as an Indian would say.

The last mails have brought two long and sociable letters from you—just the kind of letters I like to read—dashing and off-hand, and full of information. To borrow your own words: "Keep up the fire," and I'll return gun for gun.

Before I send for my bucks—(I shall not want them until next summer)—I will describe the kind of animals I require, or will describe my full-blood Merino ewes to you, and rely upon your own good judgment in making selections.

"Old Poll" I look upon as the most perfect type of a sheep for Texas—for out-door rough, work. You are mistaken in your estimate of his size and quantity of his wool. He is a stout, solid square-built, heavy sheep, sheared as high as 16 lbs. of wool in his younger days, and wool which would not shrink more than 4 lbs. in the washing, has an iron constitution, is a hearty feeder, is always fat and healthy, and for his height and length is the heaviest sheep I ever owned. I never weighed him; but I know that it takes a stout man to lift him. His legs are stout, short, and set well apart; he has tremendous hams and shoulders; he gets more lambs than any other buck, and they all look just like him, and have the same hardy qualities. You can see him all through my flocks as plain as a church, and I would not sell out his strain or blood—the element he has infused among my sheep—for thousands of dollars. He was got some eleven years ago by one of Stephen Atwood's bucks,[2] which had been purchased and sent to Vermont: he is out of a ewe owned I think by Judge Marsh.[3] For nine years he has been "roughing it" in Texas, entirely without shelter, and with no other food than what he could pick on the hill sides and in the prairie vallies: he has never had any sickness—has always been fat, strong and exceedingly vigorous—is the only one alive out of twenty-four purchased at the same time—and now looks as bright as a lamb, and as though he would live another half score of years. You must not say anything against "Old Poll," for he has been a lasting and firm friend of mine. For the qualities we require in Texas—constitution, strength, hearty appetite, adaptability to the grasses of the country, size, form, shape, every thing—I shall never own his like again[.]

*Jan. 2.*–We are having a cool, crisp air to-day; but the sun shines out brightly, and it is the finest possible weather for sheep like mine—all in good health and fine condition. But not so with new comers from Missouri and Mexico, over-driven on the roads, and arriving in October and November, especially during the latter month, in poor condition: they have not the blood, flesh, or stamina to stand up against the cold.

There is now a flock of 200 Missouri ewes in a valley in

---

[2] Stephen Atwood (1785–1867), of Woodbury, Connecticut, was a breeder of merinos for half a century. His flock of pure blood merinos was founded by a single ewe which he bought of Colonel David Humphreys in 1813 for $120. Humphreys (1748–1810) had won acclaim in New England by his importation of twenty-one rams and seventy ewes from Spain in 1802. Atwood greatly increased the average yield of his sheep.

[3] Joseph Marsh, of Hinesburgh, Vermont, bred pure blood Atwood merino sheep for about thirty years, from the mid 1840's.

front of my house, arrived a few days since, or during my absence, which I visited this morning. Around the camp I noticed three or four dead sheep, and half a dozen lambs; many of those alive are as thin as transparencies. "What do you suppose is the matter with my sheep?" queried the owner! "Starvation! and with many in the last degree!" was the only honest response I could make. The sheep have been driven, and over driven, along the roads, and where every spear of grass was cropped to the roots: how could it be otherwise than that they should starve to death. "What shall I do with them?" again queried the owner. "Take them to a spring two miles from here, turn the strong ones out to graze at sunrise, and keep them out till sunset, and give the poor ewes with lambs a little meal and water every night and morning," was my answer. Whether he will do it or not is more than I can say.

*Jan.* 3.–a lovely day, with mild southerly wind. The New Year is behaving itself quite well so far; it may soon, however, get in the tantrums, and give us another series of chilling, sleety Northers. So far, out of 10 tons of hay I had cut in November, I have not fed out 2 tons: have enough left for almost any kind of a storm, or series of storms for that matter. You shall know how I get along.

Many thanks for your kind compliments to Mrs. K.: I thought her comely when we were married, ten years and more since, and have never changed my mind. We have conferred together, and weighed well, all that you say about your youth, personal attractions, and Don Juanish ways generally, and have made up our minds to run the risk of renewing our invitation to you to pay us a visit. When may we look for you? March, when you are up to your knees in snow and slush, is a good time to start, and April and May pleasant months in Texas. Who knows but what you might become so much enchanted with our country and climate as to break up in the North, and set up your lares and penates hereabouts. Stranger things have happened.

I find I am at the end of my paper and will conclude by expression [sic] the hope that we may see you under our shelter ere long.

Truly yours, &c
Geo. Wilkins Kendall


Rancho near New Braunfels,
Jan. 9, 1860.

My Dear Sir: Just in from my sheep Estancia—left there at 9 o'clock this morning, when all my flocks were even in finer condition than ever. My two ewe flocks—now all with lamb—it would do you good to see—all bright and healthy, and as fat as seals. I have another ewe flock of some 700— last spring's lambs—which are not quite so fat: but as they are growing rapidly, this may account for it.

On the 6th inst. we had a warm, searching, soaking rain, of some ten hours' duration—a glorious rain—just the rain we needed. Since then it has been so mild and springlike, that green grass has started up in all directions. To-day I rode all the way down, (30 miles,) in a light summer coat, and was marvellously comfortable.

*Jan. 10.*–Mild again to-day—mild as May—too warm for the season—and such weather cannot possibly last: we must soon have a visitation from the Arctic regions to pay for it.

I have been all day ploughing in a little 10 acre pet field near my house—a field in which I raised nearly 50 bushels of corn to the acre, both in '58 and '59, not-withstanding the drougths [sic].

"He that by the plough would thrive,
Must either hold himself or drive." [4]

*[Franklin.*

---

[4] "He that by the Plough would thrive,
Himself must either hold or drive."
—Preface to *Poor Richard Improved* (1758).

I both held and drove to-day, using 2 horses, for I could not learn a new hired man to turn over the furrows to suit me. The ground was in such capital order, that it was fairly fun to turn it over. And then the exercise gave an extra relish to a fat, tender and juicy leg of mutton, which I brought down with me on my last trip, and which Mrs. K. cooked to a delicate turn for dinner, while a few slices of it, cold for supper, I found rather palatable than otherwise. You never sat down before a finer leg of mutton—never will until you visit the mountains of Texas. Call me prejudiced, extravagant—crazy if you will—but I know what mutton is as well as the next man, and tell you nothing but the honest truth.

From all quarters of Texas I continue to hear of frightful losses among sheep, mainly among new comers. A man named Anderson, near Gonzales,[5] lost 300 out of 800 in December—all Mexican—and this seems to be about an average loss among this class of sheep. In Mexico, as you are doubtless well aware, the bucks run with the ewes the year round, and a proportion of those brought in the past fall were of course with lamb, and unable to withstand the storms[.]

But it is not among Mexican sheep alone that losses have occurred: those brought in from Missouri and Arkansas, (and report has it that 500,000 crossed Red River the last year,) have suffered terribly, especially those over-driven, and brought in poor. Did I inform you, in my last, of one man who stacked up and burnt some 700 in a pile, all of which had died in a single night? I heard this stated as a fact, and I can well believe it to be a fact from many specimens of Missouri and other Western ewes I have seen. Many have

---

[5] Gonzales is southeast of New Braunfels, on the Guadalupe River, in Gonzales County. In February, 1854, after an unusual cold spell, Olmsted saw many dead cattle along the road to Gonzales. He remarked (A Journey through Texas, 235): "A little care to provide shelter and fodder for these rare occasions would prevent this great suffering and loss; but in not one instance did we see any such forethought."

young lambs drooping and tagging behind them—others are having lambs—and all such, both ewes and lambs, cannot withstand such storms as we have had.

While at my estancia, yesterday morning, I had a talk with my head shepherd [6] as to the kind of bucks we should require next fall for our Merino ewes. My present stock is made up of a cross of French, Silesian and Vermont Merinoes— good-sized, well formed, hardy animals—but all wanting that black, gummy appearance now so fashionable. My head shepherd has been with me now seven years, and on talking the matter over we both came to the conclusion that the black, gummy appearance soon wears itself out in Texas, where the animals are raised entirely out of doors. We recollected perfectly well that "Old Poll" himself, and all my first importations from Vermont, (24 in number,) were dark—dark as the new Merinoes now coming in—and that now they are all light. May it not be a fact that in this climate, and among sheep raised without shelter, they lose their dingy, dirty, oily, and gummy quality of wool? I shall investigate this matter. I know that my original stock of Merinoes were not troubled or treated with the "Cornwall finish," [7] which you so graphi-

---

[6] Joseph Tait, who had been hired by Kendall in Scotland in 1852. Born near the Tweed River, among the Cheviot Hills, he came from a family of shepherds.

[7] A dark external color in the American merino suggested the presence of a desirable quantity of yolk, or gum, which increased the weight of the wool, and hence was more esteemed than a light color. Sometimes fraud was resorted to in order to secure the desired effect. The passage in the *Texas Almanac for 1860* referred to by Kendall is as follows:

"Breeders defer more or less to the tastes of buyers, and thus more 'grease and wrinkles' are produced than would otherwise be. A pettier personage—your nomadic ram-peddler—carries his complaisance still further. He *manufactures* traits or peculiarities to please purchasers! He buys up half or three-quarter bred Merinos, which chance to have abundance of 'wrinkles,' (the mongrel get of a very 'wrinkly' ram often show this peculiarity quite as strongly as his full-blood descendants,) and if the natural gum is wanting, he puts it on by daubing them over, immediately after shearing, with a pigment of linseed oil

cally describe in your long and interesting article in the last Almanac.[8]

My idea is, at present, that a couple of square-built, compact, chunky bucks, of fine Spanish blood, and with constitutions so strong that the thing fairly sticks out, will do my business next fall. About wrinkles I care nothing: I have bucks wrinkled all over, from the tip of their noses to the end of their tails, and do not affect them on that account.[9] Nor do I care so much about the quality or quantity of wool; the gummy, oily element I wish to introduce if it can be done without taking away from lustiness and hardihood. There are three requisites I require:—the first, is *constitution;* the second, *constitution;* the last, *constitution.* But I will talk this matter of the bucks over with you again: my wife says I have worked enough to-day, and that it is time to go to bed. So: *bon soir* to you.

*Jan. 11.*–Here it comes, another Norther, accompanied by

---

and burnt umber—a composition known in the North as the 'Cornwall finish,' from the fact that it was first used (as a winter protection to sheep, I presume) in Cornwall, Vermont. It soon makes a nearly black external coating, so similar to the natural gum as to be entirely undistinguishable from it, except to a very practised eye. I should say, however, that it was usually a little more evenly put on, and a shade *handsomer,* than the natural article! A second good oiling, with clear oil, towards fall, helps along. Armed with these painted mongrels, a demure face, and a certificate of pedigree, purporting to be signed by a 'Deacon,' and a 'Judge of Probate,' your ram-peddler sallies forth, Macedonian-like, conquering and to conquer—greenhorns!"— Pp. 111–12.

[8] In 1857 the *Galveston News* began publication of the annual *Texas Almanac,* which was designed as a handbook of information about Texas. Both Kendall and Randall made contributions to the *Almanac.*

[9] Many merino breeders of the period sought to obtain sheep having many folds or "wrinkles" in their skin on the ground that such animals would produce heavier fleeces. Sometimes the business was overdone and the result was an animal with a skin very difficult to shear. For a description of the merino breeder's ideal in regard to wrinkles see Randall, *The Practical Shepherd,* 71.

rain which freezes as it falls—the worst kind of a storm for sheep. It came whistling and roaring along about daylight this morning, and by 10 o'clock trees, fences, and all were loaded with ice. The Northers of 2d and 6th December were colder, but were not a whit more injurious to flocks than this will be: in the former the rain froze high in air, and came down in the shape of fine hail, rolling off the sheep as it fell; now it comes down a simple rain, and immediately congeals. Thousands—yes, tens of thousands—of old ewes and young lambs must inevitably perish in such a storm: there is no earthly chance to save them. If the storm lasts two or three days I may lose some of my latest lambs and older ewes; but I have no fear of great loss.

This winter of '59–60 will prove a useful lesson, although dearly purchased, to hundreds of new owners, bitten by the sheep mania: it will teach them that they cannot buy sheep helter-skelter, drive them long distances, bring them here after frost has cut down the grass in the fall, and then expect them to "rough it" successfully through the winter.

I heard yesterday that Shaeffer [10] had arrived on the coast with his sheep: should it be true, and he is caught in this

---

[10] F. W. Shaeffer was born in Ohio, was in business in New York City for a time, and about 1857 began sheep raising in Texas, establishing himself north of San Antonio. Later he moved to Nueces County, where in time he acquired about 80,000 acres of land, enclosing the whole with a wire fence. In 1878 he had 15,000 head of sheep. He brought many Northern sheep into Texas. After Kendall's death he obtained 1,500 breeding ewes from the journalist's estate. See John L. Hayes, "Sheep Husbandry in the South," in *Bulletin of the National Association of Wool Manufacturers*, VIII, 63–75.

In the fall of 1859 Shaeffer was at Cortland, New York, to obtain sheep from Randall. He then took his sheep down the Ohio and Mississippi rivers to New Orleans, and thence by steamboat to Texas. He complained of the loss of ewes with lambs out of season, and asked Randall to make good the loss. Randall indignantly refused to do so, pointing out that the ewes were not from Randall's own flock but that they had been bought by Randall for Shaeffer at the latter's request. Shaeffer to Randall, 10 Jan. 1860; Randall to Shaeffer (draft), 24 Jan. 1860, Randall Papers.

storm on the road, away from shelter and fodder, it will go
hard with the poorer portions of his stock. But I hope that
he is not on the way up, or that if he is, the storm may soon
abate.

Two young men from Vermont, Fisk and Chase, arrived
about a month since with 104 fine Merino bucks; they halted
near Goliad [11] to rest and recruit the animals; the latter got
hold of some kind of poison wild pea or bean vines, and 70
died in a week. Hard, was it not?

*Jan. 12*–The Norther has moderated down somewhat to-
day: it is however still cloudy, thaws but little, and trees and
fences loaded with ice. Not during the winters of '57–8– and
9 was there a Norther so severe as this has been—nothing like
it. At the same time, those of 2d and 6th December were
much more severe than this has been. The winter of '59–60
will not soon be forgotten.

It has been the great misfortune with the greater number
of those who have gone into the sheep business the past year
that they knew nothing about it—hardly a sheep from a
goat—and certainly did not know the difference between
a Saxon and Southdown. Many thought they had nothing to
do but purchase a flock of sheep, put the first green-horn
who came along in charge, and their fortunes were made. If
their ewes, which they bought in July and August, were with
lamb, so much the better: they would get along the faster,
and have quicker returns. The losses of many of this class
will be well nigh total, they will become disgusted, and
will leave Texas crying that it is no country for sheep. Some,
seeing the errors they have made, may stick by and try it
again, adopting a different plan; but I repeat that many will
leave for their former homes in the Old States, and never
return. We can do without them; yet they will undoubtedly
fasten a bad name upon Texas, and retard a more legitimate
emigration.

---

[11] Goliad was a small settlement on the San Antonio River, in
Goliad County, northwest of Indianola, the port on Matagorda Bay at
which the sheep of Fisk and Chase were probably landed.

*Jan. 13*–The sun is peering out this morning, the ice is melting away slowly, it will probably all disappear by night, and I am in hopes we shall now have a week of good weather. This will give my flocks a chance to pick up all they have lost, if they have fallen off any of consequence, which I doubt. Fat and healthy sheep are not much affected if they happen to lose their feed for a day or two; but a poor old ewe, with a young lamb by her side, must succumb. I will endeavor to ascertain the effects of the storm just over, and will let you know. But we shall never know the full extent of the loss among sheep this winter: the owners, for the most part, are too proud or too sensitive to tell the whole story. "How many did you lose in December?" said I the other day, to a neighbor who bought some 1500 Mexican ewes in September. "Over one hundred," was the response. As near as I could calculate, by the look of the flock and the sign around the pens, *over* one hundred meant *nearly* five hundred.

Are you not having an uncommonly cold winter at the North?

I shall visit my sheep in a day or two, and will let you know how I am getting along.

In haste. Truly your friend,

Geo. Wilkins Kendall

Rancho near New Braunfels,
Jan. 18, 1860.

My Dear Sir:–Yours of 29th ult. came to hand last evening: am glad to hear that your Spanish ewes "tough it out" so well during the hard winter you are having. It is this element, in its pure state, that I wish to introduce. With the exception of a few Silesian and French ewes, all old, my main stock of full blood ewes have a mixture of Silesian, French and "Old Poll" in them, all mixed up. Every year I cross with my choicest bucks, and I have bred in-and-in but little.[12]

---

[12] "Breeding in-and-in is ordinarily understood, in our country, to mean breeding between relatives, without reference to the degree of

I am still busy with my ploughing: ground in fine order, and I wish to rush it through. We are all confident of good crops this next summer: I always make corn enough to do me. *I will have it.*

Three gentlemen from Boston, all prospecting for homes, spent last night with me: they left my sheep estancia yesterday morning, and one of them is negotiating for a place near me, and almost the only one now for sale. They say that my flocks were looking finely—lost two since I left on the 9th inst: one got drowned by accident—the other gave out during the sleety Norther of the 11th. This is excellent luck, as the storm was very severe. Weather now very fine.

I received a letter by last mail from a sturdy old Democrat of Sherborn, Mass. He commences as follows: "I wish to get out of this d-d abolition hole of a Massachusetts whilst I have money enough to start with: what are the chances out in Texas?" I like that man hugely, and shall do my best to get him out here. He says he is obliged to hear at least one abolition sermon every Sunday, and two prayers for the "niggers," and can't stand it any longer—the infliction is too great.

*Jan. 19*–Weather still magnificent—a warm, Southerly breeze blowing: ground in the finest possible order for ploughing, and I am as hard at it as a man well can be, and in order to have it *done right,* I am compelled to do it myself. Oh! for the *arms* of Briareos: [13] with my own single

---

consanguinity. . . . But this is not the sense in which it has been used by those eminent European writers who have done so much to plant an inveterate prejudice against its very name in the public mind. Sir John Selbright ranks among the highest of these, and he did not consider procreation between father and daughter, and mother and son, to be breeding in-and-in!***I apprehend that he means by it breeding the father with the daughter and again with the granddaughter, or the mother with the son and again with the grand-son."— Randall, *The Practical Shepherd,* 116.

[13] A mythological giant having 50 heads and 100 hands.

*head* I can get along well enough, but unfortunately I have but two arms and two hands.

I heard to-day, from a passing traveller, that the last Norther had killed many sheep below me, towards the coast. No *poor* sheep could stand such a storm, without food and shelter, and at present prices for corn and fodder any common animal would eat its head off in a week. Have not heard a word as yet from Shaeffer: but he can scuffle through if any one can.

Did I inform you that a gentleman, all the way from New Zealand, called upon me a few weeks since? When the individual from the "other side of Jordan" gets along, I shall set him down as about the "last run of shad." In sober truth, we are just now, in this poor Texas of ours, having visitations from all the *four* quarters of the world, and some even from the *fifth*. It remains to be seen whether the great losses many of the new comers will sustain this winter, (I mean in the sheep line) will cause any abatement in the fever.

*Jan. 25*–Since the 20th we have had dark, damp, foggy mornings, and pleasant days enough, and some of our grasses are struggling to put on a green dress. But they do not grow fast enough to do the sheep much good—are perhaps a detriment, as the flocks, once getting a taste, are constantly hunting for the green tufts springing up, and do not find sufficient to pay for the time and labor.

*Jan. 26.*–We have another Norther this morning, but it is moderate and dry, and will do no other harm than checking the growth of the grass. I start up to my sheep place to-morrow or next day, with a party of friends who are on a visit,[14] and will inform you how my flocks stand the weather and hard fare on my return.

---

[14] Francis Asbury Lumsden (1800–1860), with whom Kendall had started the *Picayune* in 1837, Fletcher Westray, Kendall's New York City agent, and another friend spent about three weeks with Kendall early in 1860. The guests spent much of their time hunting. While still at the ranch Lumsden wrote: "As a country for game, you have

*Feb. 3.*–Have just returned from my estancia—left there yesterday morning, and "camped" on the bosom of Mother Earth last night. The early morning hours were frosty: but my friends lived through, and came out of their blankets in good plight at day-light.

Well, I am happy to say that my flocks are still in fair condition, and all bright and healthy. The two breeding ewe flocks, and the old wether flock, are in much the best order— the most of them still fat: the two lamb flocks, (now some 8 months old [ )], have fallen off most—do not tread up as strong and as firm as I could wish. Still, unless we have a succession of severe Northers, and a cold and backward spring, I am in hopes I shall carry the most of them through. But I do not shout yet—mind that: I have got through the worst of it, yet am still in the woods. All our old weather-wises are prognosticating an early spring. Nous verrons.[15]

Since 1st December the sum total of my losses amounts to simply *seven:* one old ewe and two lambs from poverty, one lamb drowned, and three died from inflammation. In my hospital (a pasture of about 100 acres with a shed and two hay stacks in the corner,) I have about 60 sheep, old and young. In front of the shed is a rack, in which the hay is put, and every evening at sun down, as the different shep-

---

my positive assurance that no region, that I know of can surpass this. I have almost worn myself out in hunting, and there is hardly any kind of game that we do not have at every meal—three times a day. Deer are very abundant. . . . Wild geese, cranes, ducks, rabbits . . . squirrels, snipe, partridges and doves, in myriads, are everywhere to be met with. . . ."—*Daily Picayune,* 21 Feb. 1860.

Of the exploits of his friends Kendall wrote facetiously: "Such wholesale slaughter I have never witnessed. . . . Had the party re-mained here a month longer, there would have been as fine an opening in New Braunfels for a powder mill and shot tower as in any place I wot of: the amount of ammunition expended in three weeks was in-credible. The popping and banging about here, for several days, put me in mind of the siege of Vera Cruz in '47, before the big guns got into play."—*Daily Picayune,* 4 March 1860.

[15] We shall see. (Fr.)

herds bring in their flocks, they are told to catch out any animal that has lagged or drooped during the day. As the flocks range out three or four miles, the weaker animals cannot stand the work: but a week's rest in the pasture, and a pick at the hay every night, sets them up strong on their pins again, and then they are turned back into the flock. In this way I am nursing them along.

In a former letter I told you that in the fall I cut 10 tons of hay: so far, I have fed out but little over one half, and am husbanding the balance with all economy. It was good hay, cut at the right time, and salted a little in the stack: I paid $100 for it, and I believe that it has already saved me $1000, and I would not sell what is left for five times what it cost. We may not require hay again for years: but I shall always keep a store of it on hand, to feed out in sleet or raw rain storms.

*Feb.* 9.–We are having fine weather, and the grass is springing again. To-day I am comfortable by an open window, and in my shirt sleeves. I hope it may continue for a week: but there is no knowing what may come.

I have kept this poor scrawl by me longer than usual— have been very busy all the time. Yours ever

Geo. Wilkins Kendall

Rancho near New Braunfels
Feb. 22, 1860.

My Dear Sir:–I have just arrived from another visit to my sheep estancia, which I left at 9 o'clock this morning, and find yours of the 2d inst here. It has been a long time on the road, but is most welcome. It is always a pleasure, as well as a relief, to get hold of one of your off-hand, crisp, dashing, to-the-point epistles, and especially to one who is incessantly bored by a thousand and one idle, inexperienced, ignorant twaddlers, who are asking me more questions about sheep than a dozen fast pens could answer. "The books are closed":

or to come down to plain English, I have suspended—answering the swarms of anxious enquirers about Texas and sheep-raising, save by advising all and singular to come out and look at the country, and if they like it, remain—if not, go home where they came from. On looking over the batch of letters now before me, I can hardly divest myself of the idea that I am not keeping an "Intelligence Office." A young lady in Mississippi wishes me to procure her a situation as governess or teacher; a young man in Kentucky, a carpenter by trade who owns a chest of tools and an interest in a Maltese Jack,[16] wishes me to procure work for both; a lawyer in Connecticut, who says he cannot make his salt at home, expects me to find him a lucrative location here:—I might go on, but you can see how I am pestered. I repeat, then, that it is a relief to get hold of one of your famous epistles; and my wife this evening picked yours out of a dozen or more and handed it to me first, as she knew the hand-writing, and that it would interest me.

How about the sheep? you will ask. I left my flocks this morning still in fair condition, although I can see, plainly enough, that they have fallen off in flesh and strength since I last wrote you. None have died, however, and if we can only have moderate weather ahead, and an early spring, I cannot anticipate heavy losses—do not think that even twenty will die unless we have severe wet Northers. Since I last wrote you we have had mild weather in the main, and green grass is struggling its way up. But there is as yet not sufficient nourishment in it to do the sheep much good, and my great hope is that mine may hold their own for yet a fortnight or three weeks longer. Night before last we had a heavy thunder shower—a regular soaker: yesterday and to-day the weather has been almost too cool to start the grass; but as the roots are well moistened, the first warm days we have will bring it up.

---

[16] A jackass of the Maltese breed.

Could you see my range, it would be a marvel to you how my sheep have lived at all since the December storms killed the grass to the roots. For three or four miles from the pens the eye rests upon literally nothing; yet the flocks come in at night looking full enough, and were there sufficient nourishment in what they eat they would not fall off. My breeding ewe flocks have stood it altogther the best:—are in good condition. This is most fortunate, as of course they are my main stay and dependance [sic]. Whether they will be able to tough it out a fortnight or three weeks longer, without losing flesh and strength materially, remains to be seen. I hope for the best; but sheep cannot live and thrive on hope alone—grass is better.

*Feb. 23.*–A cool, bracing, blustering Norther came whistling along this morning, much to my disappointment although it is dry. The misfortune is, that it will check the growth of the grass which has been struggling out, and, if followed by a frost, will cut down what is already above ground. I trust the gale may soon blow itself out, and be followed by warm, growing weather. Our peach trees are now in bloom—so are the wild plums—while the inventor of the "Balm of a Thousand Flowers" might select quite a variety on the prairies. And here comes a Norther to spoil all! Well: there's no help for it.

You may have heard that Shaeffer had extremely bad luck with his last importation of sheep: he left some 120 head down by Yorktown,[17] and in a few days some 70 died—poisoned outright by eating some weed, or else pannicked to death by drinking bad water. The survivors he has brought up to his place near Böerne,[18] and, although they are in indifferent condition, I believe they are mending. A man named

[17] Yorktown is in Dewitt County about midway between Powderhorn, the port on Matagorda Bay where Shaeffer had landed his sheep, and New Braunfels.

[18] Kendall often used the umlaut incorrectly in writing the name of the town.

Chase, with some 80 odd fine bucks from Vermont, lost 60 near the same place a month before. Another lot of fine sheep, from Boston, suffered terribly on ship-board from over-crowding: they were from Vermont, and those still alive I am told are doing badly: they can hardly do otherwise unless well fed, and hay and fodder, as well as corn, are very scarce all over the country. Next season there will be great quantities of hay cut in Texas, and very likely no necessity to feed it out the following winter. Not the least necessity was there for hay during the winters of '57 –8 and 9. Hereafter, however, I intend to have a large quantity cut and securely housed or stacked: I shall feel more secure.

The scab [19] seems to be spreading, or at least I have heard of it recently in new localities. So far, I have escaped, and I shall do my best to guard my flocks against the disease, and should it break out I shall fight it with a will and a determination.

*Feb. 24.*–The Norther is still blowing, but fortunately it continues dry, the sky is clear, and it is not cold enough to do any other harm than to check the growth of the grass. As the present wind came in with a new moon, it may last until it changes: we shall see.

I have no earthly objection to your using such portions of my letters as may serve, trusting to your own good judgment in selecting such parts as cannot compromise me with some of my neighbors. As yet I have not seen your last letter in the Galveston News, but am anxiously looking for it.

So soon as the Norther moderates I shall commence planting corn: I have already put in beets, carrots, cabbages, turnips, peas and other "sass" in my garden, and as the ground is moist and in fine order I hope to have a good show

---

[19] Scab, a disease resulting in loss of wool and sometimes death, is caused by an animal parasite which fastens itself to the skin of the sheep or burrows into it, irritating the tissues and producing an effusion.

of early vegetables. In the long run, I have heretofore had fair average luck with my garden.

I will write you again so soon as any thing turns up which may in any way interest you.

Truly your friend,<br>Geo. Wilkins Kendall

P.S. Excuse the oil spot above: it got on the paper by accident, and I have no time to re-write.

Rancho near New Braunfels,<br>Feb. 26, 1860.

My Dear Sir:—I finished a letter to you two days since, and took it down to the post office. The Norther I then spoke of lulled down before I got back, the night was pleasant, and yesterday it was as mild as May. To-day we have a fidgetty, unstable, raw, disagreeable wind, baffling about from S.E. to S.W. with some slight signs of rain. I hope, however, that it will hold up for a few days, as I wish to plant my corn immediately. My ground is now in capital order, and the quicker I get my seed in the better. If a frost comes, well: I can plant again. If it does not come, I shall get a good start with my crop, and a shower or two in March and April will make it past all peradventure.

One of my Teutonic neighbors [20] brought up the mail from town this evening, and among a raft of letters I found yours of the 8th inst. You are at perfect liberty to use any thing I have written you about our weather this winter, or any thing I have said about my flocks, and the way they have so far weathered the storms since December set in: I only ask you to so alter any of my hasty scribbling that it may appear in good English, for I have little doubt I may have sinned

---

[20] Kendall was living in a section whose population was predominantly German. German migrants founded New Braunfels in 1845.

against Lindley Murray [21]—committed many murders upon the Queen's and our vernacular. I am compelled to write at what your Northern fast men would call a 2:40 lick,[22] or not get through my correspondence at all: a heavy correspondence it is, I can assure you.

In my last I told you all about my flocks—how they were getting along, &c, &c. Having a chance yesterday to send a note to my head shepherd, I told him, in case the weather turned cool and he saw the least sign of farther falling off in the two lamb flocks, to move them down to a fresh range four miles south of my estancia, and give them the chance of fresh picking. He will do so if he sees the least necessity for it. Without boasting, I believe my sheep are now in better condition than any others within the wide limits of Texas, taking size of flocks or numbers into consideration, and if I can worry along for a fortnight I am all safe—barring disease. Of this I am a little fearful, although I have no other reason to be apprehensive than that I have escaped it so long, and my sheep have been so universally healthy since May, 1856. The children are all in bed, it is 9 o'clock, my wife admonishes me that I have much to do to-morrow, (says I must plant a lot of peas among other things,) so I will close for the night. Weather getting more mild, and the air a little damp: I hope it may not rain to-morrow.

*Feb. 27.–Evening.–* We have had a lovely day—warm and cloudy—just the right kind of a day for the young grass. I have been hard at work, ever since breakfast: have well nigh finished my garden, and have commenced planting in my main or principal field. You must know, or you shall know, that I am excessively nervous: I wish every thing

---

[21] Lindley Murray (1745–1826) was an American grammarian whose works were widely used and who for many years was looked upon as the ultimate authority on questions of grammar. It has been said (*Dictionary of American Biography*, XIII (1935), 365) that "he was to grammar what Hoyle was to whist."

[22] A racing term, indicating a mile covered by a trotter in 2 minutes and 40 seconds.

done in a particular way, and in order to have it so done I am compelled to take hold myself. To-day, among other amusements, I have taken a hand at ploughing, dropping, and covering corn—covering with a hoe, in good old New Hampshire style—and I am extremely happy to say that every thing, so far, has been done to suit me. If I dont [sic] make a good crop, it will be no fault of mine.

I note all that you say about the bucks, and soon as you have sheared I wish you to select for me a couple of full-blood Spanish Merinoes—compact, hardy, vigorous spec-imens—build them a house (or pen) send them to New York at such time as a friend of mine there, Mr. Fletcher Westray, may designate, who will ship them to New Orleans by steamer. From thence they will be taken, by first boat, to Indianola,[23] where I shall either be myself, or else have a mule team in readiness to bring them up. Should I change my plan in any particular hereafter, I will let you know in due season. A leggy, slab-sided, gangling, over-grown, hot-house, forced up, monster of a sheep, I hold in utter con-tempt, no matter what the quantity or quality of wool: my Merino ewes, at present, are a mixed up mess of Silesian (Geo. Campbell,)[24] French, (of my own choice and im-portation,) and Old Poll, (the latter the noblest Roman of them all, by a long ways,). The pure Spanish element I wish to introduce, and I trust entirely to your own good

---

[23] On the west shore of Matagorda Bay, in Calhoun County, a few miles southeast of its rival, Port Lavaca. During the year ending 31 Aug. 1860, 2,561 sheep were included among the imports at Indianola. Over five million dollars worth of exports, it was estimated, including 1,412 bales of wool, left the port during the same period. See *Texas Almanac for 1861*, 242.

A new wharf, 2,100 feet long, was built at Indianola in 1860. The port was finally destroyed by storms in 1886.

[24] George Campbell was a breeder of merinos at Westminster, Ver-mont. Silesian merinos were descendants of Spanish merinos which had been taken to Silesia during the Napoleonic period. They were introduced into the United States during the 1850's, George Camp-bell being one of the first purchasers.

judgment in making the selection. My old hour, 9 o'clock, has come again; I have a hard stent [sic] before me tomorrow: so, good night.

*Feb. 28,*–Whoora! for Texas: we had a glorious shower during the night, accompanied by good solid thunder and lightning, and it is at it again this morning, in real downright earnest. We have now had quite enough of a good thing to last us a spell, and I hope it may hold up raining long enough for me to get my corn in. Of course it is too wet to continue planting this morning. We have had a sufficiency of rain to keep our main springs running nearly all summer —an important matter. The weather has been warm and muggy all day, and I can almost see the grass grow: a few such days, and we shall have a sufficiency for all our stock.

I do honestly think I had the best leg of mutton on my table to-day I ever tasted—fat, tender, juicy, fine-flavored, delicious. I brought it down with me six days ago, and it was just right.

In my next hope to tell you how my sheep are doing.

> Yours truly,
> Geo. Wilkins Kendall

> Rancho near New Braunfels
> Friday eve, March 9, '60

My Dear Sir: I can see my way pretty well out of the woods now: left my sheep estancia at 9 o'clock this morning: all my flocks improving fast: the lambs set to skipping and jumping, as playful as porpoises, when they turn them out of a morning—they are now nearly a year old, and getting fat every day. Shall be out of the woods in a fortnight,—will not the [sic] ring the bell until I am away out in the open.

We have had gorgeous weather until yesterday, when a straggling norther came along. But it was mild and dry, and had a good effect in checking the too rapid growth of the grass. The entire face of the country is now as green as a wheat field—trees all leaving out.

Total number of deaths in my flocks, since 1st November, exactly fourteen:–five puny sheep killed by wolves; three drowned; two killed by accident; and four died from cold or poverty, call it what you will. What do you think of that statement?

Can't you drop in and dine with me the day after to-morrow, Sunday? I started this morning with the fattest kind of a wether in my wagon: on the way down saw seven deer looking quietly at me from a mat of live oaks: singled out a fat young buck, raised my rifle, and knocked him over. Tied my horses, dressed him, and put him in the wagon. We shall have venison and mutton on Sunday, and room for you at table.

Good night.

Truly yours

Geo. Wilkins Kendall

*March 10.*–Magnificent weather to-day: corn coming up.

Rancho near New Braunfels

March 19, 1860.

My Dear Sir:–Just in from my sheep estancia: left there at 8 o'clock this morning, when all my flocks were in the best possible condition: yearling lambs picking up and gaining fast. No more deaths: sum total of losses since 1st December, from cold and disease, *six!* This is not much for a hard winter, eh? What do you think of it?

Weather has been dry and coolish since I last wrote, and sky uncommonly smoky. Growth of grass has been checked, but still there is sufficient for all the flocks. Had the weather been wet and warm it would have grown too fast—would have been brack and watery. A good shower just now would be advantageous; but the signs are dry, and the croakers are again growling.

On 20th of last October I put my full-blood Merino ewes to buck, and this morning we had seven fresh animals. All look promising enough, although the ewes would have been in better condition, and given more milk, a week or ten days hence. But as they are improving every day, I think the lambs will all "make the trip." The first comer was a buck, from a pure Silesian ewe, and he is a rouser. I shall watch him. He was got by an old but favorite Vermont buck.

*March 22.*–The weather continues cool and dry, and a warm shower is more and more needed. Still my corn, which is all up, is not suffering, and the grass is so good that all our stock here at the rancho—horses and cattle—continues to thrive. My brood mares, which have run out all winter without shelter, are getting fat every day: a lot came up this evening, after their salt, (I herd altogether with salt,) were as fat as seals. Not a hoof, either of horses or cattle have I lost this winter, which says something for our mountain region. How all have "toughed it out," and "held their own," is to me a marvel. Not a sign of food have they had save what they could crop. But just about here we have deep ravines and cedar brakes, affording fine shelter against Northers, and to this fact I mainly owe my success. On flat prairies a three days Norther will run the stock out of all reach or hearing.

You may have heard something of the profits of stock-raising in Texas, or I should say horse raising: let me give you an idea from my own experience. Six years ago, last August, I purchased, among a lot of others, a Mexican mare, with a filly colt by her side, for $11: last spring the old mare, (including herself,) had a family of *fourteen,* and there would have been *fifteen* had not a 2 yr old stud colt died from altering. This spring I am expecting an addition of four more to the family, which will make eighteen in all. And what have they cost me? Simply the salt they have used whenever they come up after it, and the trouble of brand-

ing. Not one of them would know an ear of corn from the
Erie Canal, were they side and side, and in seven years I
can safely say that the old mare has earned me, clear of all
expense $700, for I would not sell her progeny for that sum
to-day. But dont publish this: it might set some one crazy,
and there are madmen enough already. For a Mexican mare,
the old one is a magnificent specimen, and no one need expect such luck as I have had. At the same time with [out]
arrogating too much, I will say that few will watch as close
and work as hard: this a perverse, stiff-necked and *lazy*
generation.

*March 24.*–We had a smart little shower about 1 o'clock
this morning: it would have been a heavy pour had not a
norther set in and dispersed the cloud. It tried to rain again
this forenoon, but was too windy—a raw and cold northerly
wind.

Five months ago to-morrow I put 22 bucks with some 1100
ewes, and I presume the lambs will commence coming to-
day. It will be hard on some of the new comers, if the
weather is as bad at the estancia as it is here: but unless it
rains to-night I shall not lose many. Our corn and grass need
a heavy rain: our lambs need dry, warm weather. We cannot have everything to suit us.

*After noon.*–It has faired off warm and bright, and the
lambs may now come as fast as they please. On 1st November
I put 21 bucks with some 950 ewes: in the course of a week
I shall be looking for nearly 200 lambs a day. I may have
mentioned these facts before:–if so, excuse repetition.

*March 25.*–Fine, pleasant weather to-day: the light shower
has freshened the grass somewhat, and cleared the air. I
shall visit the estancia in two or three days, and on my return will write you again—will inform you how this year's
lambs look. They will average nearly ⅞ Merino.[25]

Truly yours,
Geo. Wilkins Kendall

[25] Kendall had been steadily improving his flock by the use of pure
blood merino rams. The offspring of such a ram and a Mexican ewe

Rancho near New Braunfels,
March 26, 1860.

My Dear Sir:—I sent a boy down to the village yesterday
with a letter to you: he brought back yours of the 9th inst.,
containing sample of wool from "Honest" buck. He—but
before I say another word, let me describe the weather at
this particular writing. "Ave Maria Purissima! a las dies en
la mañana, y mucho frio!" [26] slightly to change what the
watchmen used to tell us of a night, during our campaigns
in Mexico.

When I sent my letter off yesterday forenoon, it was warm
and cloudy, and threatening rain: at 3 in the afternoon it
tried to rain, but could only sift down a small sprinkle: at 4
a dry norther set in, and continued until our usual bed time
at the rancho, 9 o'clock. This morning, at day-light, I was
awakened by a sharp peltering on the roof of my gallery,
and on looking out found that the shingles were being pep-
pered by a storm of sleet about the size of peas! Fine times,
eh? for young lambs out in the open! and I probably had
fifty dropped last night! and they are even now coming at
the rate of six, eight, or perhaps ten every hour! for I have
often noticed that the worse the weather the faster they are
sure to come.

And then, here is my corn! Yesterday morning gales as
from Araby the Blest were softly fanning it: now, blasts
from Spitzbergen, or the regions where Sir John Franklin
went under,[27] are spitting upon it with all their fury! Well:

---

would have one-half merino blood. The offspring of a ewe with one-
half merino blood and a pure blood merino ram would have three-
quarters merino blood. The offspring of a ewe with three-quarters
merino blood and a pure blood merino ram would have seven-eighths
merino blood. As this process went on, the resulting wool would be
finer and more abundant. This was a practice of which Randall highly
approved and which he had advised.

[26] "Hail, Mary Most Pure! Ten o'clock in the morning, and very
cold!"

[27] Sir John Franklin (1786–1847) started out in 1845 on his third

I shall not cry nor fret. If it is all killed, I shall hitch up my horses and plant again. This A.D. 1860, with the help it had from 1859, is sorely trying us, but, like Mark Tapley, I half fancy that it is jolly.[28] If I am to fight, give me an industrious, open, and ever-active enemy, and then if I gain the battle the victory is so much the more satisfactory. The great charm of life with me has ever been to contend against adverse circumstances, unforeseen obstacles, and contrary running cards of all kinds. Could we all back our carts up to a gold mine, and shovel in to our hearts' content, some few lazy, worthless drones might be made happy; I should be the most miserable devil on the face of the earth. I was once dining with a friend in Paris when, on looking over the evening Galignani, I saw that there had been a great fire in Camp-street, New Orleans.[29] "The Picayune office has gone, certainly," said I. "I'll bet you a dinner it isn't," said my friend. "Done," quoth I. The next mail brought the news that we had lost $35,000 in 40 minutes! I won the dinner, with all the trimmings, an excellent dinner it was, and I was never happier in my life, albeit my loss was heavier, by about 184,925 francs, than that of my friend. During the last

---

polar expedition, which was lost in the Arctic. During the 1840's and 1850's many expeditions in search of him and his companions were sent out; it was not until 1859, however, that the story of the ill-fated expedition became known. See *Dictionary of National Biography*, XX (1889), 191–96.

[28] Mark Tapley is a character in Charles Dickens' *Martin Chuzzlewit* (published in book form in 1844, the year that saw the appearance of Kendall's book, *The Texan Santa Fe Expedition*). Mark sought to be "jolly" under the most trying conditions, and met his greatest test in the United States where he and Martin lived for a time.

[29] During Kendall's stay in France John and William Galignani were publishing in Paris *Galignani's Messenger*, an English language newspaper which their father had founded in 1814.

The fire alluded to occurred 16 Feb. 1850, destroying 22 buildings including the *Picayune* quarters at 66 Camp Street. The paper moved back to its old quarters at 72 Camp Street and never missed a morning issue. About six months later it was once more at 66 Camp, in a new four-story building.

half a century I have made it a point not to shed a tear over a drop of spilt milk: how do you like my philosophy? I have already, this morning, covered all my growing tomatoes and other tender vegetables: I know that my head shepherd, the trusty Scotchman, Joe Tait, will do all he can for the lambs; I feel that I can do no more: I am content. Good night.

*March 27.*–The Norther still continues, raw and cold, but fortunately the sleet has ceased falling: this is a great help to the lambs. My corn this morning has a sorry look: whether it is killed to the roots or not, I cannot say. Now about the bucks.

I have taken a fancy to the "Honest Ram" you describe, and think he will "fill my bill" to use a common expression about here. I care but little about the gum—that will all wear out in a year or two, and the gummiest buck in your State cannot get gummy lambs in this, if they "run out" the year round. As for wrinkles, I also care but little about them: I have bucks in my flocks the wrinkliest rascals you ever saw. I have also finer wooled bucks—much finer—some of the best Silesians ever raised by George Campbell. That slight wave in the line of his back—that rise at the withers—is the only real objection, to my poor thinking, you urge against him, while his legs, chest, vigor and constitution more than make amends for that fault with me. You should distinctly understand that I am not breeding for fancy sheep, to sell at fancy prices: my great object is to produce numbers, and to effect this I must have lustiness and constitution. Show me a buck wrinkled from nose to tail, coated with a fine and heavy fleece of the blackest and gummiest wool ever handled, and if I do not like his chest, legs, barrel and other points he shall [not] have a showing among my fine or full-blood ewes. Book the "Honest Ram" for me.

I shall probably have some 125, or perhaps 140, full blood ewes, and shall require another buck, and as both will be compelled to "rough it" with the rest of the sheep, of course I require a strong one. For the novelty of the thing, I would prefer a pure black Spaniard—as black and gummy as possible—but this must not interfere with his form and hardihood. Would not a 2 year old this spring be preferable? He would be 2 years, 6 months old when put to ewes in the fall, and is not an 18 mos. old buck almost too young? Under shelter and regularly fed, I know that Northern bucks are much larger than mine at 18 mos.—more forward in every way—and I never put my bucks to ewe until they are 2 yr years [sic] old: I have even kept some of my choice rams in the wether flock until they were 3 yr., fancying I could obtain more vigor and greater longevity in the offspring. By keeping the old buck *a la* stallion, he could easily serve 100 or 115 of my full blood ewes, and the balance I could put a younger buck next fall without injuring him. I merely throw out these suggestions, and shall depend, meantime, entirely on your own good judgment.

You speak of sending out four bucks. Of course I shall require but two for my own use next fall: the others I will sell for you if possible: but I am a little fearful the market may not be as good this year as it would have been had we had any thing like a decent winter. The losses have been terrible, all over the State, since last December; many I know have become disgusted or disheartened; others may be deterred from commencing; and the demand for all kinds of sheep materially lessened. Had I lost six *hundred* sheep the past winter, instead of simply *six,* or even *sixteen hundred,* it would have annoyed but not discouraged me; but others are not so obstinate—mulish if you will—as I am: are disposed to give up with every little run of bad luck. You know the world.

As regards shipping the bucks, I have a friend in New York, Mr. Fletcher Westray, (firm of Westray, Gibbes & Hardcastle,) No. 1, William street, who will see that they are well bestowed on one of the schooners running regularly to Lavaca or Indianola, and at the latter I have friends who will receive and send them up to me in a wagon drawn by mules. I must ask you to write to Mr. Westray, detailing the quantity and quality of food necessary for a voyage of at least 50 days: the hay or oats left over will not be lost, if not all wasted before the vessel reaches Indianola. Here's the rub.

I dislike much to risk bucks in the hands of third parties—more especially animals of great value. Could I be along myself, I should have no fear: but over-feeding, neglect of feeding, negligence, carelessness, and what not I have almost invariably found to be the rule, while attention and watchful care have been the exceptions in nine cases out of ten when sheep have been brought to Texas. Before another year rolls round I shall probably pay you a visit: I hope to be able to start in October on a hurried trip to your region, when I shall take charge of a buck, and perhaps a few ewes, all the way back to my own doors: and they will come safe if I succeed in making the trip myself, and the sheep I shall bring will be good ones: I shall coax you out of your best.

Many thanks for your kind wish that you could send me 50 tons of hay by telegraph; you will be glad to know that it was not needed. The 10 tons I cut last fall not only served to feed my work oxen and horses, but all my sick and enfeebled old sheep, and such lambs as could not make a daily trip out of four miles and back. But take my word for it, if grass grows this summer, I shall have at least 40 tons of hay cut, and much of this I can store away under the roof of my stone sheep shed, which is 150 feet long by 16 wide, and well shingled.

I note all you say about barns, and entirely approve your suggestions. But you must know that for the last four years, or during my entire administration of the sheep business, I have been sorely troubled with what the French term "une faute d'argent," or what we sometimes call, a shortness of money, and have not been able to make all the improvements I could wish. My plan has been, from the first, not to sell any ewe so long as I thought she would produce, and no wether so long as he could earn me from 25 cents to $1.00 per annum from his back. The man who makes a crop of wheat, cotton or sugar sends it to market in the fall, sells it, and pockets the produce: the man who makes a crop of lambs, if he follows my system, adds them to his productive capital, and is eternally cramped for money. I had an offer of $1000 last fall for 100 grade 2 yr old ewes: the price was a good one: I wanted the money: but had not the nerve to sell. And why? I reasoned in this way: the

Ewes will shear 500 lbs wool, at 25c. per lb.—$125.00
Ewes will give 90 lambs, worth $5.00 each   450.00
Total——   $575.00

A tolerably good interest on $1000. You must understand, as a matter of course, that it cost me nothing to keep the ewes, save the salt, as I could not have reduced the number of my shepherds.

I have written this off at race horse speed, and am fearful you cannot decypher it. But you will perhaps understand the drift of it.

At this time, (3 o'clock P.M. [)] the sun is struggling out, and the norther is moderating. In a day or two I shall visit my sheep and will write you immediately on my return.

Truly your friend<br>Geo. Wilkins Kendall

I will ask Mr. Westray to inform you the exact time when a schooner sails for Indianola [.]

Rancho near New Braunfels,<br>April 2, 1860.

My Dear Sir:–I have only time to say that I came down from my sheep estancia last evening, and that yesterday morning I left over 200 fine lambs, the oldest but eight days, and that the ewes are giving an abundance of milk. By the end of this week I shall be able to count 1200 lambs, and in a fortnight or three weeks close upon 1800; and as the ewes are in tip-top condition, I hope to save nearly all. Could you be along about 1st of June, and see them all jumping and playing, the sight would be rather pleasant than otherwise.

During the cold sleet storm of Monday last, the 26th ult. (about which I wrote you,) I only lost six lambs, and had a surplus of twins to give one to each of the old ewes which lost her lamb. Since Monday the weather has been coolish but dry; to-day it is a little warmer, and I have some hope we shall now have a shower, now much needed for the crop.

Promising you a longer letter next time, I subscribe, In great haste,

Your friend, truly,
Geo. Wilkins Kendall

Rancho near New Braunfels,
April 6, 1860.

My Dear Sir:–Your letter dated plainly on the 1st March, but which was as plainly post-marked on the 17th, came to hand last evening. By some accident it has got mislaid, and I cannot just now lay hands on it; yet I recollect the contents, and will commence an answer.

You ask information more particularly about the practicability of a route for sheep from some point on the Mississippi or upper Red River to the San Antonio region, and on this question I am sorry to say that I cannot advise you. Once on the Sabine, and across it, provided you had decent weather, you would be all safe.

You recollect of course what John Randolph [30] said about

---

[30] John Randolph of Roanoke (1773–1833) was a well-known Virginia politician during the early days of the Republic. He was celebrated for his brilliance, his eccentricities and the sharpness of his wit. He once declared that he would go a mile out of his way any day to kick a sheep.

the Ohio—that it was frozen up one half the year and dried up the other half: I found it so too often in early times for either my temper or comfort. Red River, too, is very low in summer, at least it has been so some seasons, and you might get shut out there and driven to take land at Gaines' Landing.[31] Of the route through from there to the Texas line I know nothing, nor am I acquainted with any one of whom I could ask information. It is often travelled by persons settling in Northeastern Texas, but I have never seen a man who has made the trip.

I presume I have been asked a hundred times, by persons wishing to commence wool growing in Western Texas, how they could get a flock through from Tennessee, Kentucky, or Ohio? and I have been unable to answer them. It is now thirty years since I went up Red River, and fifteen since I was in the interior of Arkansas, and in those days I was not thinking about sheep. I know I have that I have [sic] seen the swamps of Arkansas dry—and again I have seen them so fat that they would bog a blanket. I know, too that there are many poisonous weeds in some of the swamps both of Arkansas and Missouri, and that many sheep have died on the way from eating them.

Were you sure of a good stage of water in the Ohio and Red River in July or August, I have always thought I should adopt the following plan in bringing out sheep: I would have built, if practicable, a two-story flat boat, (in your case at Pittsburg) large enough to hold all my sheep comfortably, and with room enough to feed the sheep on each floor or story. I would then charter a stern-wheel steamer to tow my ark down, and put my feed on board of her. If possible, I would obtain freight to help pay expenses. I would come up Red River as far as possible, or to the best and direct

[31] Gaines' Landing, on a bend in the Mississippi River in Chicot County, Arkansas, about thirty-five miles north of the southern boundary of the state, was being used as a "jumping-off point" for sheep in 1860.

point to disembark, and then strike into Texas across the Sabine. Such is the rough outline of a plan I have always had in my head: to ascertain its entire feasibility to my satisfaction, I should probably make a reconnoissance of the route, from one end to the other.

Were I sure of a dry season, and good weather, I think I could drive a flock through Arkansas, starting from say Gaines' Landing on the Mississippi; but I can assert nothing from my own experience, nor from that of any one who has ever brought sheep through by that route. It is some 7 or 800 miles from here to the Mississippi: I now forget the exact distance.

*April 7.*–We are still having dry, windy, but pleasant weather enough—excellent weather for lambs just coming. My corn holds its own tolerably well: we need a warm rain, however, for grass, for gardens, &c, &c., and our corn will soon require a shower. If our crops fail this year in Western Texas, the last of those engaged mostly in farming will be driven East, and all the old corn and cotton fields will be turned into pastures.

To go back to sheep. Fancy yourself caught in August on the Ohio, on board a steamboat drawing three feet *large,* with only two feet water *scant* on the bars, and where are you? Many and many a time I have been caught in that fix; later in the season, it is true, but still caught—sparring and jumping over bars, and every moment expecting to be blown into mince-meat. If you send a flock by that route, be careful to ascertain, beforehand, that there is plenty of water.

A well-constructed flat, roofed over, would sell for nearly as much as it would cost, on the Mississippi or Red River, to break up for the timber and lumber: at least such is my impression. To the mouth of Red River it could be towed easily if there was sufficient water: up Red River, if the water was low, would be the rub. Why, I have spent a week on a sand bar, on board a boat which the captain said "could run any where where it was moist," and how would your sheep fare similarly situated? Look out for these things, and above all things, have a careful, industrious, watchful man with the flock.

A man named Peel, who has purchased a place near me, has now gone to Tennessee and North Mississippi to purchase a flock of sheep to drive through—at least such was his intention if, after going over the ground, he thought it would be feasible, or prudent. If I can hear any thing of him, I will at once inform you. Or if I can gain any information in relation to the route through from Gaines' Landing to the Texas line, I will communicate it immediately. I have frequently mentioned Gaines' Landing, on the Mississippi, as a jumping-off point: there may be a better one higher up or lower down. As I never affected that region, I never took much notice of it.

The name of my agent in New York is Fletcher Westray, of the firm of Westray, Gibbes & Hardcastle, No. 1, William street. I will ask Mr. Westray to inform you when a good schooner is about to sail for Indianola; in season for you [to] have the proper pen made for the bucks, and to send them to New York. I look upon it as a risky business—sending them out without a friend or protector—but cannot do otherwise. I think you will see me at Cortland Village, and with Mrs. K. should it not be sickly in New Orleans, some time in October. If so, I shall constitute myself an escort to a buck, and perhaps half a dozen ewes, and as my time is valuable I shall want the best.

Truly yours,
Geo. Wilkins Kendall

On giving my papers a general jail delivery this morning, I found your letter.

I will write you again soon, and especially about route for sheep if I can learn any thing.

Rancho near New Braunfels
May 10, 1860.

My Dear Sir:–I have been remiss in writing to you of late: cause, I have had little to say and much to do. So soon as the lambing season was well over we were in the midst of shearing; my crops have required attention constantly; and my brood mares (running out,) I have been obliged to hunt up and bring home to the stallion. So, you must readily understand I have had my hands full.

I left my sheep estancia on the 7th inst.—I then had a little over 1800 lambs, and they are altogether an improvement upon the drop of any previous season—stouter, fatter, stronger, better shaped and more vigorous—not a drooper among them. One buck lamb, in particular, comes nearer my idea of perfection than any thing of the sheep kind I ever laid eyes upon—the most hypercritical could not find a fault in him. His mother was got by "Old Poll" out of one of Geo. Campbell's Silesian ewes; his father was got by a favorite French buck of mine out of another of Geo. Campbell's Silesian ewes: such is his pedigree. He is wrinkled from the end of his nose to the tip of his tail, has the chest of a prize-fighter, is as straight in the back as an arrow, is square-built, compact, and as heavy as lead, stands firmly on four short but wide set and substantial legs, and walks like a drum major. And then he will be a "muley" or hornless buck, like his grandsire, "Old Poll," which gives him an additional value in my eyes. For five or six years, hand-running, I visited the Imperial French flock, at Rambouillet,[32] during

---

[32] Spanish merinos imported into France in the 18th century were the foundation of the famous government flock at Rambouillet. Over the years the French, at Rambouillet and elsewhere, produced a distinct type of merino which was known as the French merino or the Rambouillet. Although some merinos were brought from France to the United States early in the nineteenth century, the significant importation of French merinos did not begin until 1840. Although they were larger than American merinos and yielded more wool, early importations were also less robust. Randall estimated in 1859 that "between

or immediately after lambing time, but never saw as fine a specimen there; and if Baron D'Aurier,[33] the director, could see such an animal, money would not buy him. Come out and look at him, and tell me candidly what you think of my judgment.

My wethers, this year, out-sheared my calculations: many gave a clip of 8½ and 9 lbs., one sheared 11 and another 12 lbs., the wool entirely free from burrs and grass seed of all kinds, and not at all oily or gummy to give it extra weight. In Texas, I am confident, we cannot look for oil, gum, or any thing calculated to increase the weight of fleece as at the North, nor do I believe that my wool would shrink a quarter in the washing. I shall try the experiment thoroughly next year, should I live, as I intend moving to my estancia with my family this summer, and shall have abundant leisure on my hands.

Although many of my yearlings sheared 6, 7 and 8 lbs this year, I do not think the entire flock averaged more than 4½ lbs when I looked for 5 all round. My head shepherd accounts for this by saying that they (the lambs) did not get enough to eat in January and February, when they were growing and required more nourishment than the grown wethers. This may be so.

My clip this year will come up to 18,000 lbs. To show you the increase during my administration of 4 years—(I came here to live permanently in May '56,) read the following:–

eighty and ninety per cent of our best Northern and Eastern wool-growers" then thoroughly detested French merinos; he himself thought that the reaction had gone too far.

Further breeding in France produced greater vigor and more wool. During the 1880's and 1890's importations from the flock of Baron von Homeyer of Pomerania began the modern history of the Rambouillet in the United States. William G. Markham, in whose home near Avon, New York, the Randall Papers were found, exhibited a number of the baron's sheep at the Columbian Exposition in 1893, arousing great interest.

[33] Jean Baptiste Auguste, Baron Daurier (1804–1869).

| In May, | 1856, | I sheared | 2800 | lbs. |
|---|---|---|---|---|
| " | " 1857 | " " | 5000 | " |
| " | " 1858 | " " | 8500 | " |
| " | " 1859 | " " | 12,000 | " |
| " | " 1860 | " " | 18,000 | " |

This, considering that I have been constantly selling off, and not adding save by natural increase, will do.

We have had dry and cool weather in the main since I last wrote—excellent weather for ripening the wheat crops, which are heavy about here, but not so good for the corn. I now have roasting ears, however, in my garden, my main field is tasseling out, and if we can have a single shower while it is in silk I shall have a heavy yield. All depends upon a rain within the coming ten days or a fortnight.

*May 11.*–I have read Lord Macauley's [sic] letter to you— have read it twice—and regret to say that I believe every word of it.[34] But other causes than those he mentions will help to split us up and dissever our Union: indolence, extravagance, corruption, and an increasing desire to live fatly without toil are all digging away at the vitals of our Republic: a want of all respect for the constituted laws of the land is constantly increasing: new "isms" are every day

---

[34] Thomas Babington Macaulay (1800–1859) wrote a letter to Henry S. Randall on 23 May 1857, in which he said: "I have long been convinced that institutions purely democratic must, sooner or later, destroy liberty, or civilization, or both." The historian predicted that democracy in the United States would, in the twentieth century, as population grew and economic conditions worsened, produce either dictatorship or ruin. "Thinking thus," he wrote, "of course, I cannot reckon Jefferson among the benefactors of mankind." The publication of this letter after Macaulay's death raised the question whether it accurately represented the historian's views; some doubt even was cast upon its authenticity. A letter from Randall on the subject appeared in the *New York Times*, 12 June 1860. Randall presented excerpts from other letters which he had received from Macaulay to show that the views expressed in the letter of 23 May 1857 were not merely views of the moment. For Macaulay's letters and their story, see "What Did Macaulay Say about America?" in *Bulletin of the New York Public Library*, July, 1925, 459–81.

starting up: vice and immorality are every day crowding honesty and virtue from their homes: legislation has got to be but another name for swindling: while a huge standing army of pestilent politicians of all parties, far more productive of evil than the martial hosts of France, Austria or Russia, is constantly clamoring for place and power. The honest, conservative element is still strong in our midst, undoubtedly; but I doubt whether it can always contend in pitched battles with the rampant spirit of faction, which is ever industrious in its dirty work. But a truce to politics: I shall lose my temper if I go on.

By to-day's mail I have written a letter to Mr. Fletcher Westray, my friend in New York, requesting him to inform you when a vessel is up for Indianola, her day of sailing, &c. &c. Mr. Westray will see that the bucks you send him for me will be well bestowed on board. Be kind enough to inform him as to the quantity and quality of food necessary for a voyage of 40 days at least. Any hay or oats left over can be sold to advantage at Indianola, the oats at least. I shall have the bucks brought up from Indianola by wagon (mule team) [.] You will also be kind enough to have a proper pen constructed for the bucks while on board the vessel, and be explicit in all your directions to Mr. Westray, and he will do all in his power to see the bucks safely sent off.[35] I presume this will reach you about shearing time.

Now that lambing and shearing time is over, I will endeavor to write a little oftener—at least when any thing transpires which may interest you.

Truly your friend,<br>Geo. Wilkins Kendall

New comers, from all quarters, are flocking in by scores, all prospecting for sheep places or homes in the mountains.

---

[35] Westray apparently did not enjoy his assignment. On 20 July 1860 he wrote to Randall: "This is the last operation I want to have with sheep for account of any one." Randall Papers.

Rancho near New Braunfels,
May 15, '60.

My Dear Sir:—Just in from another visit to my sheep
estancia, and find here a letter from D. Richardson,[36] of the
Galveston News, wishing me to hurry an article I have promised for the Almanac of 1861, and also to furnish you, at my
earliest convenience, a "series of data on climatology," which
he thinks I have promised you. As I have kept no diary or
record of the kind, of course this will be out of my power:
I trust you have all my hastily written letters by you, as they
will give you a species of rough record of our weather during the fall of '59 and winter of '60. In these letters I have
described, I believe, every norther, every thunder shower,
every storm of wind or rain, we have had, and I trust they
will answer your purpose. I am confident I have given you
the duration and force of all our worst storms, whether dry
or accompanied by rain or sleet, and the list is formidable.

Our dry spell continues; it has hardly rained a drop for
three weeks, and my corn, now in tassel and beginning to
silk, is commencing to suffer. If it does not rain soon—say
within a week—I shall have but an indifferent yield: a single
good shower, any time between this and the 21th [sic] (new
moon on the 20th) would give me an abundant harvest.
The weather has however been favorable to wheat, rye and
barley: I have three or four acres of the latter heading out
heavily, and next October, or early in November, intend
sowing or ploughing in 20 or 30 acres. If the ground is in
good order then (wet enough,) it will furnish an excellent
winter pasture for sick or disabled sheep—my little patch
served a good purpose last winter. I also intend ploughing

---

[36] David Richardson (c. 1815–1871) joined the Galveston News in
1852. In 1860 he was associated with Willard Richardson (1802–
1875) in publishing the News and the Texas Almanac; he was planning to move to New York City. He and Randall were personally
acquainted. For an account of the Almanac see Stuart McGregor,
"The Texas Almanac, 1857–1873," in Southwestern Historical Quarterly, L (April, 1947), 419–30.

in some 10 acres of winter rye and wheat next fall, as pasturage, with the hope of making a good crop in the spring. We must all turn our attention to small winter grain hereabouts.

I left my flock in the same splendid condition—all fat and doing finely. The grass in Blanco county, notwithstanding the dry weather, continues excellent. My only fear is that our spring, which has now been running constantly for 8 years to my certain knowledge, will give out unless fed by copious showers, and then—what? Why, I must commence a wandering or nomadic life with a vengeance, and go prowling about the country, like a Tartar, for water and grass. But I will hope for better things—will continue in the belief that we shall have rain enough "to do us" before summer is over.

Did I inform you, in my last, that I have hope that my this year's clip of wool will be sold in Georgia, and manufactured there? [37] It is not yet "un fait accompli," but I think I shall not only succeed in introducing my own wool into the greatest of our Southern manufacturing States, and at a rate which will satisfy the consumers, but open a road in that direction for all or at least much of our Texas wool, and if I do, I shall be just as happy as Douglas would have been had he received the nomination at Charleston on first ballot.[38] In New York there is no chance for us to get our rights. What with forwarding charges, freights, commissions, guarantees, *and the cry against Texas wool* among the buyers of that city, we never obtain our own.[39]

*May 16.*–We had a slight shower before day light this

[37] Eleven small establishments producing woolen goods were enumerated in Georgia by the census of 1860. The value of their combined product was less than half a million dollars.

[38] The Democratic National Convention had met at Charleston the month before this letter was written, with Henry S. Randall a delegate from New York.

[39] Much of the Texas wool was produced on Mexican sheep or on sheep resulting from a cross with Mexican ewes. It was frequently coarse and very poorly prepared for market. Kendall himself was a strong advocate of grading fleeces and bagging them carefully.

morning, but not enough to effect much good. There probably would have been a heavy rain had it not been for a strong, blustering Northeast wind, which dispersed the clouds to the four quarters of heaven. The sprinkle, however, freshened the corn and grass a little, and will help considerably if we can soon have another and heavier shower. Old residents hereabouts prognosticate a wet summer: nous verrons, as old Father Ritchie used to say. It is high time we should have one wet season.

I will drop you a line whenever any thing turns up of the least interest.

Truly your friend,<br>Geo. Wilkins Kendall

Rancho near New Braunfels,<br>June 2, 1860.

My Dear Sir:–I have just received a note from D. Richardson, of the News, enclosing your long and most interesting article on "Sheep-Raising in Texas," cut from the columns of that paper. Mr. R. says that if I have any alterations to suggest I can write you immediately, so that you can make such alterations before the letters are stereotyped for the coming Almanac.[40]

I read the letters as they appeared in the News—I have read them again—and can suggest no alteration of the least importance save in the very first paragraph: "Lambing." You there say, after speaking of the necessity of running some risks at the North in having early lambs, that "In Texas you have no such occasion for haste, and farmers who are unprovided with shelters may as well wait a little and be on

---

[40] Randall's article, entitled "Summer and Winter Management of Sheep in Texas," appeared in *The Texas Almanac for 1861* (Galveston, 1860), 152–66. The previous year he had published in *The Texas Almanac for 1860*, 102–16, a more general article entitled "Sheep Husbandry in Texas."

the safe side." Now, here you are in error: we cannot wait
without running a great risk of having stunted, pot-bellied,
spindle-shanked lambs. I left my sheep estancia yesterday
morning, and saw the fact "sticking out" as plain as a church
steeple. If the month of May were moderately cool and
pleasant, with an occasional shower of rain to keep the grass
sweet and tender, a lamb dropped say as late as 20th of April
could get along tolerably well; but it is seldom we have such
weather in May. On the contrary it is hot and dry, and the
later lambs show it almost from the start.

My lambs this year commenced dropping in earnest about
the 25th of March, and by 10th of April, as near as I can
now recollect, I had close upon 1500 out of the 1800 I can
now count. By 25th of April all had come, or at least I do
not think half a dozen came after that date. I am now coming to the point, which is, that a lamb dropped on the 1st of
April will be a heavy, lusty and almost a made sheep when
he is two months old, while one dropped even on the 20th
of April will "hang fire," become puny and pot-bellied, and
look like an entirely different animal. My head shepherd
yesterday was talking about picking out some 150 or 200 of
the late comers, and putting them in a pasture kept tolerably
fresh, simply because they were falling off, while he pointed
out at least 800 of the first comers which were compact and
square-built, and nearly as large as their dams. I believe that
a lamb dropped on 1st of April will be as heavy when three
months old as a lamb dropped on 1st of May will be when
six months old, while the former will stand altogether a better chance of going through his first winter in condition,
and consequently in safety. The later lambs get caught by
hot and dry weather, and coarse and tough grass, get
stunted, and always keep behind. We this year had fine
growing showers between 18th and 25th of April: we have
hardly had a drop of rain since, and the grass, although still
good enough for the sheep and older lambs, is too hard for
the younger. And so it is three years out of four, and hence

the necessity of having early lambs, and the risk of putting off weaning time. Could we work it so that all our lambs could come "in a pile," say on 25th of March or 1st April, we could raise all. I would commence having lambs on 10th of March, were I not fearful of being caught in a *wet* Norther: a dry one is nothing. No weather affects my young lambs after they are dry, on their feet, and commence milking—at least nothing in the shape of cold: but a hot, dry day, in May or June on a lamb two or three weeks old, is terribly trying.

For June lambs, in ordinary years, I would not give a dime apiece.

We have had no rain for five or six weeks, and our corn crop will be short: much of the wheat and rye gathered, and a good yield. I shall experiment with barley ( *winter* barley) this fall: shall sow some 30 or 40 acres for a winter pasture, believing I can have a good crop in the spring of grain.

I have scratched this off in the greatest haste to save a mail: excuse repetition, omissions, &c. &c. Will write again soon.

Truly your friend,<br>Geo. Wilkins Kendall

Rancho near New Braunfels,<br>June 17, 1860.

My Dear Sir:–Yesterday's mail brought yours of the 3d inst. I was wondering what had become of you—was almost fearful that they had used you up with their fine old Madeira at Charleston—am happy to learn that you lived through the siege, and can readily imagine you found lots of work before you on your return home.[41]

Here in Texas we are having an exceedingly hot and dry

---

[41] "I have just returned from the city of Charleston—eleven hundred miles from here—& what with the fatigue of travel & what with the reaction after exertion & excitement, I feel like a man 200 years old." Randall to Hugh Blair Grigsby, 12 May 1860, in Frank J. and Frank

summer. In the mountains about my estancia we have had two or three good showers, giving us fair crops of wheat, rye and corn, (of the latter I have a fine crop;) but below, towards the coast, they have had little or no rain, and corn will be very scarce and dear. Grass, too, is burnt out in many localities on the lower prairies, and stock must suffer unless it rains soon. With us in the mountains all our animals are rolling in fat, and could they get hold of one of our 3 or 4 year old steers in New York, they would cover him over with rosettes and ribbons, precede him by a band of music, and show him through the streets as a prize beef. So with our horses—all are in fine order and fat.

I left my sheep estancia day before yesterday: all well. A few of the later lambs, perhaps 50 or 75, are not quite up to the mark, but I can show you at least 1700 beauties—round, fat and sleek, and nearly as tall as their dams. More than ever am I convinced of the necessity of having our lambs come early in April—the long, hot, dry days of May and June curl up the late comers. Could I make a business of it, and have all my lambs come on the 1st of April, it would be all the better. Next fall I shall have more bucks with my ewes than ever, and not allow them to remain with them later than 15th November: I am determined. And any ewe which does not take between 20th October and 15th November, may run barren for a year, and rest herself.

I note all that you say about the Ugly Buck, and trust that ere this he is on the way out, with his mate. But if he has his coat on, and it continues as warm as at present, he will melt. Yet I would much like to see him with his fleece on, and should he reach Indianolo unshorn he will have cool and safe conduct up to the "Rancho near New Braunfels." I already have a conveyance engaged.

A neighbor of mine purchased a buck of Patterson [42] last

W. Klingberg, eds., *The Correspondence between Henry Stephens Randall and Hugh Blair Grigsby, 1856–1861.*

[42] John D. Patterson of Westfield, New York, a leading breeder of French merinos, many of which he sent to Texas. He later engaged in

fall, a Frenchman, brought him here, and kept him about the house. He was a fine specimen, shapely enough, and really sheared, about 1st of May, 25 lbs. of wool in the dirt —the growth of God only knows how many months or years. The other night he got out of the yard, and when found the wolves had taken ac't of stock of his carcase! My neighbor has some 20 ewes with lamb by him; but all will come in the fall, and will be of little account unless hot-house raised.

I should not have the least objection should you send out two extra bucks to me, or two dozen for that matter, provided I could sell them at fair prices. But I have been dubious, all the time, as to the market the coming fall—have been fearful that the heavy losses incurred last winter would lower the prices of sheep, and prevent the sale of fine stock. I am now inclined to believe, however, that the sheep fever will increase, and that where one person has become disheartened or disgusted, two new comers will be ready to step in his shoes. My own good success the past winter, and that of some of my neighbors (old residents,)—still looms up as proof that we have a good sheep country.

A word about the article for Richardson's Almanac for '61. I have already written and sent on mine,[43] and have only made two points: in the first, I have mentioned my own great success the past winter, and given the causes; in the second, I have alluded to the bad luck of many of the new comers, and have offered my poor opinions as to the reasons —not a word about climatology, farther than that I have said the past was an unprecedently cold and severe winter.

In my letters to you, written last winter, I think that I described every change of weather which happened: it certainly was my intention, and I trust you now have the letters

sheep raising in California and Texas. Randall secured information from Patterson when he was writing *The Practical Shepherd*—see page 36 of that book.

[43] Kendall's article, "Sheep-Raising in Texas," which appeared in *The Texas Almanac for 1861*, 166–70, was the fourth report on sheep raising which he had made for the *Almanac*; his eighth and last one appeared in *The Texas Almanac for 1867*, 217–19.

by you, and if so, you might add a paragraph to your article, simply giving the number, length, and date of the different Northers happening from the 12th of November, when we had our first cold snap and frost. The mild weather which set in about middle of February, and which started the grass a little, was what carried me through: had the Northers continued later, it would have gone hard with my flocks.

I believe you have never suggested to me the propriety of writing about the weather of last winter. I could have done so, had I thought of it; but it is too late now, and you must take hold of the matter yourself if you think it worth while.

If you have them by you, please send the letters by two of Lord Macaulay's former secretaries, as well as your reply: I am most anxious to see the dressing you will give the gentlemen.[44]

I wish you could be here just now, and spend a month with me: I'll give you a programme of what we would do. (Get your map of Texas open.) After resting a few days at my rancho here, we would start up for my estancia near Böerne in a Jersey wagon,[45] well provided with blankets, coffee, pipes, hams, bread, and Bourbon whiskey made by a friend of mine before the invention of strychnine. We would kill a deer and "camp" on the route. I would insure you against snakes and rheumatics, and you would say you never knew before what pure air and refreshing sleep was. We would remain two or three days at the estancia, where I could show you a buck lamb, (grandson of "Old Poll,") which comes as near perfection as you will often see. We would next go to Campe Verde[46] by way of Comfort[47]—capital

---

[44] See note 34, above.

[45] A flat-bottomed wagon with large wheels and cloth-covered top, used to carry passengers and freight; called after the state where many of them were made.

[46] Camp Verde was a United States army post on Verde Creek, in Kerr County, northwest of New Braunfels; it was established in 1856.

[47] A German settlement on the upper Guadalupe, in 1860 a part of Kerr County, today a part of Kendall County.

fellows on the route, and lovely country. At Camp Verde we would spend a day, (the U.S. officers there always glad to see gentlemen,) look at the breeding camels, and have a good time generally. A Turk, Arab, or some sort of a turbaned rascal told me last year that a 2 year old camel raised in Texas was larger than a 3 year old at home, and better formed. I am no judge of this kind of stock, and cannot say.[48]

We would next drop over through Bandera Pass [49] to Bandera City,[50] and examine the country on the upper Medina and its tributaries—a lovely region which is now attracting much attention. (Mr. Heyward, a gentleman with charming wife and sisters, recently settled there, and others moving in.) We would take a look at the "Mound league" as it is called, (has a mound in the centre overlooking some 5000 acres hemmed in by smooth and grassy mountains—price $1.50 per acre—next year it will be $3.00 if not $5.00.) Next to the rancho of an old friend of mine on the San Geronimo—a prince in his way, who would give us as fine

[48] The United States government brought some 75 camels to this country in 1856 and 1857 in an effort to improve transportation in the West. As Secretary of War Jefferson Davis was keenly interested in the experiment. Although the camels proved adaptable, the government after the Civil War sold at auction those that remained. During this period some camels were also imported by private individuals.

In a letter appearing in the *Picayune* on 6 July 1856, Kendall told of a visit which he had made to San Antonio in June. He wrote: "Just as I entered, the camels with their Arab attendants, were coming in, causing a general excitement among the population, and a general stampede among all the horses within sight of the strange procession. It is not every town in the new world that can boast of having witnessed such a scene, and my own mind was carried away to Cairo and other cities of the East, where a caravan of some forty camels is nothing to stare it. The last I saw of the animals they were browsing about among the mesquit trees near the San Pedro Springs, looking patient, contented, and apparently well reconciled to their new home."

[49] Bandera Pass, about three miles from the site of Camp Verde, is a natural gorge about 125 feet in width through the mountains separating the Medina and Guadalupe river valleys.

[50] A settlement on the Medina River, the county seat of Bandera County.

roast kid and fried bass as you ever tasted. Our next move would be to San Antonio, and thence back here, and I am willing to take oath that you would say you had never breathed purer air, seen clearer water, or set eyes upon a more lovely country since you were born upon earth.

"Dos't like the picture?" as Claude Melnotte said to Pauline Deschapelles, after describing to her the section around some pond in the old country, vulgarly called the Lake of Como.[51] When are you coming out?

I find, in running over this random scrawl, that I have called the Honest Ram the Ugly Buck: I presume I was thinking of your Plug Ugly buck in my haste. He must be on the way by this time, unless the vessels running to Matagorda Bay have hauled up. I believe, however, that one or more packet schooners are always running in the trade. This section is still overrun with strangers, prospecting for homes.

I will write you again whenever any thing of interest turns up.

Truly your friend,<br>Geo. Wilkins Kendall

I believe that Shaeffer leaves to-day for Mexico with a lot of bucks. I have not seen him for some time. He has sold all his sheep.

Rancho near New Braunfels,<br>July 7, '60.

My Dear Sir:–On coming down from my Sheep Estancia yesterday I found your long and most interesting letter of 20th ult., with slip from the N. Y. Times,[52] on my table. While dissenting from you in many of your views, I still

---

[51] Claude Melnotte and Pauline Deschapelles are characters in Edward Bulwer-Lytton's play, *The Lady of Lyons* (1838). The Lake of Como is a beautifully situated lake in northern Italy.

[52] Probably containing Randall's letter on his correspondence with Macaulay, which appeared in the *Times* on 12 June 1860.

fervently trust that your hopefulness as regards the durability of our Republic may be verified. If we can rub through the coming four years from 4th of next March, and hold together, we may go on; but there's the pinch. Some day when I have leisure I will give you my views as to the future of our country: I have no time at present to say more than that the tone and handling of your letter in the Times pleased me much.

We are having a fearfully hot and dry time of it in Texas: from New Braunfels and other points at the foot of the mountains, all the way to Matagorda Bay, the country is a parched and burnt district. No grass, no water—no teams running—and if my bucks are now on the way, and arrive at Indianola before a searching rain is vouchsafed us, I hardly know how I shall get them safely up. Here at my rancho, four miles from the foot of the hills, I have made a splendid crop of corn—much of it ripe, and we are now eating new bread: below, on nine acres out of ten, they will hardly make seed. Western Texas, so far certainly as the planting interests are concerned, will get a terrible set-back this year, and one from which she will not easily recover.

Meanwhile, in the mountains higher up, and all the way to Fredericksburg,[53] the crops of corn are good, and the grass still holds out: at my own Estancia, yesterday morning, it was as green as spring time, and my sheep and lambs never looked better. So far I have been most fortunate—how long my good luck will last I cannot say. But my grass will certainly hold out until August, and if my spring continues to run for six weeks longer, I shall begin to look for our regular fall rains. And should they come, we shall have excellent fall grass, and my flocks will get a good send-off for the winter storms. Will they come mild next winter? or shall we have a repetition of '59–60? Nos veremos.[54]

How is the weather with you this summer? Here it is

---

[53] A German settlement in Gillespie County, northwest of New Braunfels.

[54] We shall see. (Sp.)

scorching—absolutely burning. Down in town to-day, in a shady place, the thermometer stood at 105!!! Texan history makes no mention of any thing like it. It is not so outrageously hot up in the hills; yet quite warm enough, and for four days past my time has been spent out in an open wagon, in the broiling sun! And my nights? you may ask. Why, on the bosom of Mother Earth, and, God be praised, our nights here are always cool, giving one a chance to recuperate.

On the morning of the 4th I was in San Antonio, with my wife and one of my children: to get rid of the noise, dust, and hullabaloo we started up to our sheep place, 26 miles North. Arrived there, we took lodgings in a hundred acre pasture—our room a corner of the fence, our bed a fresh plot of grass. And not a man in your empire State of New York, not a woman or child, ever breathed air so pure, or had sleep so refreshing, as was ours—never will. Laugh at me if you will, but don't pity me: I don't need it.

To-morrow morning, at 3 o'clock, I start up again for the Estancia, to meet a builder: I am going to work at once to put up shelter for all my household, and shall move up in the course of six weeks or two months. There will always be a room for you there, under good cypress shingles.*

I have said that we "put up" in a corner of the fence on the night of the 4th July. Yes, and it took two French blankets —none of your flimsy American trash—to keep us comfortable before midnight, and no dew falling! "How many people have got bad headache this day, in America," said Mrs. K. to me, as the sun and all of us rose in the morning, "and our heads are so bright and clear." "Certainly" said I, as we started out to see the lamb flocks.

Don't call this any thing but a hasty scrawl.

Truly your friend,
Geo. Wilkins Kendall

* You must know that as yet we have no other than shepherds houses at my Estancia, and my wife and self always "camp out[.]"

I received 28¼ c. per lb. for my wool, cash, this year, in the dirt, all round: about 5 c. more than I could have realized North.[55]

Have no time to look over this—excuse mistakes.

New Braunfels, July 19, 1860.

My Dear Sir:–Just in from my sheep Estancia in Blanco county, and find your short note of 30th June: regret exceedingly to learn that the bucks were to be shipped at New York by the 6th inst., but there is no help for it. Had I for a moment supposed that the drouth was to continue, the grass to burn out, and the springs to dry up, between this and Indianola, I should have asked you to detain the animals until you heard we had had a good, searching soaking shower; as it is, I must do the best I can.

I shall write to-day to my agents in Indianola to care for the bucks on their arrival, to water and feed them in some stable, and to write me the moment the vessel nears the wharf. If but two come, I shall start down in a Jersey wagon, pack corn and fodder, and travel through o'nights: I know the country and the short cuts. I shall take in hay and oats, and hurry back—lying over during each day in the shade. But if four or five animals come, I must hire a mule wagon and a trusty man, and send down for them cost what it may. You might as well try to drive sheep across the Desert of Sahara in mid-summer as over the route between this and Matagorda Bay just now. Why, I was told the other day that good drinking water was 50 cents a bucket full at Lavaca, and I suppose it is as dear at Indianola! No rain there for three months! On the route up, the first water is at Victoria,[56] 40 miles—a

---

[55] At this rate he would have realized a little more than $5,000 for his clip.

[56] Victoria, on the lower Guadalupe, in Victoria County, was described by Olmsted as "a very ditto of Gonzales, and all the rest. It stands on the great flat coast prairie, near the edge of the river bot-

hard journey with the thermometer at from 115 to 130 in the sun, and ranging at 103 to 105 in the shade! Such blistering weather as we have had this summer was never felt before in Texas—it would be insupportable on the Coast of Guinea. All of to-day I have been facing a perfect sirocco in an open wagon, the hot air coming ten times heated over the parched lower plains. Three weeks ago I paid a man $1.00 per 100 lbs. for bringing up a load of salt for my sheep: to-day he would not stir a step after another load for $10.00 a 100—no, not for $20.00. The trip would inevitably involve the loss of every ox in his team. So it goes. I still hope that we may have a rain before the bucks reach the coast; but if not, I repeat that I will do the best I can to save them. I hear that cattle and stock of all kinds are choking to death and starving to death in many places below; but I shall endeavor to run the gauntlet, and the verb "to fail" is not in my vocabulary when I can manage matters with my own head and hands, and have average luck.

"And how are you getting along in the mountains?" you may well ask. Most truthfully I can answer, magnificently— never better! In fact, I may say, never so well! My yearling lamb flock, nearly 1400 in number, looks better than ever before at this season: my buck and wether flock, about 1300, is a sight which, if you could set eyes upon it, would gladden you. My breeding ewes are in capital condition, and my this spring's lambs, save about 100 of the last comers, are nearly as tall as their dams, and fat as seals. The springs at my Estancia, cool and delicious, still run boldly, the grass is yet growing on the lower grounds, and even my milk cows are waddling in fat, and sleek with an abundance of food meet for them. And coming down this afternoon, seeing a caval- lada of horses some three or four hundred yards from the road, I turned to see if a fine ¾ blood mare, with a colt, might not be among them—(I had missed her for three

tom."—A *Journey through Texas,* 240. A large part of its population was German.

months.) And there she was, fat as one of Barnum's prize babys [sic],[57] and her colt in the same condition: you never set eyes upon handsomer animals since you was born upon earth. And to close, I have raised a fine crop of corn at both my places, and all that I now want to carry me swimmingly through is a single good rain, any time between this and the 1st September:–I can stand it until then, so far as my stock is concerned:–would that all the people of Texas could say the same.

This is a running, private letter—written as fast as Flora Temple's best time.[58] Excuse blunders.

Yours ever
Geo. Wilkins Kendall

I brought down a saddle of mutton—(now hanging in my well,) which I would be glad to send you. Fat, juicy tender.

Rancho near New Braunfels,
July 30, '60.

My Dear Sir:–Yours of the 10th inst. just to hand: I presume that the five bucks are now somewhere "Off the Tortugas," [59] if the vessel sailed on the 18th or 20th inst.

[57] All Americans were familiar with the name of Phineas T. Barnum (1810–1891) in 1860; his American Museum in New York City has been amusing and amazing its patrons for years. The promotion of baby contests was only one of the many activities which the showman carried on. In 1851 Kendall and his wife saw Barnum drive through the streets of New Orleans with Jenny Lind, then on her American concert tour.

[58] A famous trotter of the day, "the bob-tailed mare" of song and story. On five occasions she set the world's record; in October, 1859, at Kalamazoo, Michigan, she did the mile in 2:19 3/4. See John Hervey, *The American Trotter* (New York, 1947), 455–57.

[59] The Dry Tortugas are keys at the entrance of the Gulf of Mexico, 120 miles southwest of the southern extremity of the Florida mainland.

Rancho near New Braunfels
July 30, '60.

My Dear Sir:— yours of the 10th
inst. just to hand: I presume
that the fine bucks are now
somewhere "Off the Tortugas,"
if the vessel sailed on the 18th
or 20th inst.

I have just hired a trusty
man to go down to Indianola
after them, with a mule team
He will start the moment I
hear of their arrival: he will
take no other load, and rush
them up in seven days.

"God tempers the wind to the
shorn lamb," says Sterne. If
the wind is not tempered a
little cooler for the unshorn
"Himant Ram" than it has
been of late, he will melt—
all care shall be taken of
the animals.

I am off immediately for
my Sheep Estancia: will
write you at length as soon
as I have a leisure half
hour. I owe owing you
a long letter.

I'm in great haste,
Truly yours,
Geo. Wilkins Kendall.

I have just hired a trusty man to go down to Indianola after them, with a mule team. He will start the moment I hear of their arrival: he will take no other load, and rush them up in seven days.

"God tempers the wind to the shorn lamb," says Sterne.[60] If the wind is not tempered a little cooler for the unshorn "Honest Ram" than it has been of late, he will melt. All care shall be taken of the animals.

I am off immediately for my Sheep Estancia: will write you at length so soon as I have a leisure half hour. I am owing you a long letter.

In great haste,

Truly yours,
Geo. Wilkins Kendall

P.S. Capt. Nelson, of Bosque, I do not know—at least, I do not reccollect [sic] him.[61] Mr. Julian I have seen several times, and take him for a quiet gentleman. Parrish I have known well for some seven or eight years—a reliable man in every way—one "who will do to tie to," in the common but not very classical parlance of the West.

Weather continues hot and dry—all my stock still doing well.

Yours, &c.
G.W.K.

---

[60] Laurence Sterne, A Sentimental Journey Through France and Italy (1768).

[61] "Captain Allison Nelson, of Bosque County, Texas, visited me in 1860. He had started to bring me a pair of these Mexican dogs, but unfortunately permitted himself 'to be laughed out of it'—his friends being under the impression that it would be carrying coals to New Castle to take sheep dogs to a region where the Scotch colley was to be found in abundance."—Randall, The Practical Shepherd, 405.

The Parrish referred to is doubtless W. D. Parrish, who was living near Boerne in 1869, and was then described as "standing at the head of the sheep interest" in that section of Texas.

Rancho near New Braunfels,
Aug. 5, 1860.

My Dear Sir:–I have a breathing spell of half an hour, and devote it to you.

On the evening of the 30th ult. I was in town, and found yours of the 16th stating that the six bucks had started, at the post office. As I was about leaving for my rancho here (you should know that I reside in the mountains 4 miles above New Braunfels,) a gentleman of the Teutonic persuasion came rushing up with the gratifying intelligence that my sheep estancia near Böerne was burnt up! "How did you learn the news?" queried I, in no enviable mood. "Here is a man who passed the place yesterday morning, and saw the prairies all ablaze!" was the agreeable response. On questioning the man, he told a perfectly plausible and straight story: said that he was in Böerne two days before, saw a heavy smoke, asked where the fire was, and was told that it was at Major Kendall's sheep place. The next morning, while on his way to San Antonio, he saw the same fire burning, exactly in the direction of the estancia—saw all this with his own eyes, mind you.

The next morning, after a hasty breakfast by candle light, I was on the road up, not in remarkably good spirits, cogitating as to whither I should move my flocks. On reaching a high hill four miles this side of the estancia I could see that the prairie vallies two miles ahead of me had escaped. This was consoling. Three miles farther on, I got sight of the S.E. corner of my land, and still no signs of a fire. This was farther gratifying. On driving up, and making inquiry, I found that not a spear of grass on my land had been scorched, nor on some 15,000 acres adjoining on which my flocks graze, and that all the sheep were in tip-top condition—never better. You may readily imagine that my mind was now a little easier than it had been for the previous four-and-twenty hours: what could have induced the man in New Braunfels to start such a story is more than I can say. It may possibly be that he mistook a prairie fire on the other side of the Guadalupe as one raging on my place.

On the 1st inst. I counted all my sheep, placed a new head shepherd over them, and since then have been looking over accounts and settling with my old shepherd, Mr. Tait. He leaves me for no other reason than that he thinks he can do better by going into business for himself.[62]

The weather continues excessively dry, although it is not quite so hot as it was in July. So far both grass and water have held out at my place; but nearly all over Western Texas both have failed, and stock is suffering terribly. God grant we may soon have a searching rain.

*Aug. 6.*–I have received a letter from Mr. Westray, dated 20th July, stating that the six bucks had been shipped, and enclosing a bill of lading. By the latter, I see that the entire charges, (freight, fodder, oats, potatoes, primage,[63] &c. &c) amounts to $76.75 to Indianola: from the latter place to my rancho here I have made a contract to have all brought up for $40 in a covered wagon, (mule team,) and from this up to the estancia will cost about $10 more—or say $125 in all from New York to their final destination, or a trifle over $20 a head. This seems high to me, but there is no help for it. The passage money from N.Y. to Indianola was simply $4.50 a head, or $27.00 for the lot: the "incidentals" count. If you hate commerce any worse than I do, it is passing strange. The charges, primage, &c. amount to $49.75 from N.Y. to Indianola. But I repeat there is no help for it.

I have written to my agents at Indianola to take especial good care of the bucks on arrival, and to send me word immediately, when a man will start down for them. You must know that the entire country between this and Indianola is

---

[62] Tait apparently located near Kendall; when Indian troubles came to Kendall in 1862, Tait came to help hunt for the missing shepherds.

[63] A small payment to the ship's captain to insure careful attention to the sheep.

burnt up—not a blade of green grass, and water very scarce. Could I have anticipated all this, I should have written you in June not to send the bucks until September; but I could not look for such a withering drouth. Few or no teams are now on the road: the man who will go down for the bucks will travel nights, take corn, fodder, and a barrel of water for his mules, and bring up no other load, again travelling nights should the days continue hot. So we have to work it. I sincerely hope all the animals may come through safe, and once here I will see that they are well bestowed. I cannot but feel sanguine, although many despond, that we shall have rains in September and October—good, soaking rains —and if so, I shall have some 40 acres of winter barley and wheat in the ground for a winter pasture. I am selling off my old ewes and puny lambs to the last animal—in nautical parlance, am clearing my decks for action—and shall be better prepared this season than ever, *if it will only rain.* There's the rub.

On the arrival of the bucks I shall make my selection from the five yearlings, (for which accept my best thanks) and shall do the best I can with the balance. If they come in indifferent condition, I shall not pretend that they are for sale: if there is no rain between this and October, it will probably be the wiser course to hold them until another year, taking the best possible care of them during the coming winter. Come what may, I shall with pleasure do all I can for your best interests.

As yet I cannot form any opinion as to what the feeling will be in relation to sheep this fall: every thing will depend on the weather during the coming six or eight weeks. Look where you will now in Western Texas, save in the neighborhood of Böerne and perhaps Fredericksburg, and you see nothing but desolation and despondency; but don't for a moment think that I partake of the general gloom—quite the contrary. I shall fight my guns both larboard and starboard, bow and stern, to the very last, and so long as I can find a blade of even dry grass and a drop of water, shall not give up the ship. Nil desperandum.[64]

I shall visit my estancia again in a day or two, and will write on my return.

Truly your friend,
Geo. Wilkins Kendall

I can never thank you sufficiently for your kindness in sending me so many bucks from which to make a selection.

Rancho near New Braunfels,
Tuesday evening, Aug. 7, 1860.

My dear Sir: Jubilate!
Last evening I rode down town, and put a letter in the post office for you, along with some half a dozen others. It was exceedingly dry, windy and dusty. At one of the taverns a lot of outsiders—some of them from York State—were cussing the country generally, and themselves particularly, for ever having come to it—called themselves d-d asses, and paid themselves kindred compliments, for having set foot in Texas.

One of them, a New Yorker, purchased a place in the mountains in June last, when all was green and smiling, and in July bought a flock of sheep. Last week he sold out, "stock, lock and flute" as the saying is, and without losing any thing by the way—(my God! how it is pouring on the roof at this writing, ¼ to 9. P.M.)—and said he was going back to New York as fast as stage and steam would let him.

Just then a man came in who offered to sell his entire stock of cattle at $4.00 a head—(the common price hereabouts is about $7.50 a head [)]. "Do you know what I would do? said I, "if I had $100,000 to invest?" "No," put

---

[64] Nothing must be despaired of. (Lat.)

in the gentleman from Gotham, "what would you do with such a pile?" "Lay it out, down to the bottom dollar, in cattle at $4.00 a head, and take the chances of an early rain and a fresh start of grass," retorted I, with pointed and positive emphasis. And thus saying, I lit my meerschaum, jumped into my Jersey wagon, and rattled out of New Braunfels, thinking, as I was wending my way up the mountain which overhangs the town, that the fool-killer does not perform half the work he did when I was a boy, now some forty odd years gone by, for the man laughed at me.

Hozanna!

❖    ❖    ❖    ❖    ❖    ❖    ❖    ❖    ❖

This morning, a little after daylight, Mrs. K., as is her wont, got up and roused the servants. On coming back, she remarked that it looked uncommonly red and angry in the East. Up I jumped, looked out towards the rising sun, and saw at once that it was coming: I could smell the rain afar —farther than Job's war horse did the battle.

> "Red sky in the morning,
> Sailors take warning."
>                          [*Anonymous.*

And there was a peculiar redness about it this morning that told me the thirsty earth was about to take a drink.

I hurried on my clothes, sent a boy out after a horse, had him harnessed and hitched to a plough, and was soon hard at work in my corn field turning furrows in the old rows. You are aware that I have already made one good crop of corn this season: I am now endeavoring to raise another.* I may not succeed; but as Miller said at Lundy's Lane, "I'll try." [65]

By 8 o'clock this morning it began to rain, and before 9 I

---

[65] James Miller (1776–1851) commanded the 21st infantry at the battle of Lundy's Lane in the War of 1812. When asked whether he could take a British battery placed on a hill, his reply was, "I will try." He took the battery, and was later presented by Congress with a gold medal inscribed with his celebrated response.

was soaked to the skin. Yet I stuck to it as long as the plough would run, and afterwards planted a few rows of corn. At intervals, all day long, we have had glorious showers—searching, soaking showers: it is still pouring: the ground is saturated: in three days the prairies, yesterday as bare as billiard table, will be green as a wheat-field: our horned cattle, (they have suffered most,) will soon be rolling in fat: we shall have a good chance to make hay: and these showers, if they have been general, will be worth almost millions to Western Texas.

Let us sing!

*    *    *    *    *    *    *    *    *

In trying for a second crop of corn I am planting the Early Canada. I may not make much bread, but certainly shall have fodder. How it is pouring! Good night.

*Aug. 8.*–It kept on pouring much of the night, and is still threatening more to-day. I have been hard at it, all the forenoon, putting in corn and garden sass. Some of my neighbors are laughing at me for planting corn at this late hour: let them laugh. *If* I should happen to raise any thing like a crop, and with a single rain in September and no frost until 1st November I shall do it, my sneering friends will be mighty apt to hear of it.

Most gorgeous picking the bucks will find when they reach here: better than they ever were used to up your way. I am expecting every day to hear of their arrival at Indianola. Will keep you advised of every thing.

Here (afternoon) comes another black cloud, with just rain enough in it to keep vegetation moving at a 2:40 stride. Sunshine and showers: there *is* balm in Texas. I'm going down town this evening with the hope of seeing that New Yorker: I shall say nothing to him—I am not revengeful, nor will I ever strike a man when he is down. The person who offered to sell his cattle last evening at $4.00 a head would not take $7.00 for them to-day.

I shall plant more corn to-morrow, and next day probably go up to the estancia. Will write on my return.

Truly your friend,
Geo. Wilkins Kendall

* Corn now worth $2.75 per bushel in San Antonio, and at such prices it is an object to raise as much as one can.

Rancho near New Braunfels,
Aug. 26, '60.

My Dear Sir:--"The rain it raineth every day," [66] and every night for that matter: it commenced showering last Tuesday, and is still at it to-day, (Sunday.) We have had enough, and more than enough, of a good thing to last us for three months: who says it never rains in Texas? I am looking for an unusually wet winter: if my expectations are realized, we shall be able to test thoroughly whether this is a good sheep country or not. Many persons have been holding back, waiting to see the effect of a long, wet winter, before commencing wool growing: if I carry my flocks safely through such a winter, they will all wish they had taken hold sooner.

My second trial of corn, planted on the 8th, 9th and 10th inst. is now nearly knee high, so green that it is nearly black, and growing as if by magic. I have planted it in furrows directly between the old rows, where a fine crop of made corn is now standing, and *if* I succeed in making a second, I shall sound the loud timbrel—you will hear of it. A week of warm sunshine, and a liberal use of sweep and hoe, will force it up wonderfully.

Our prairies are now clothed in a richer green than you ever saw, and we shall be able to cut more hay this fall than ever. I have a 1280 acre tract lying unoccupied where, by

---

[66] "For the rain it raineth every day."—Shakespeare, *Twelfth Night*, Act V.

1st of October, the black or oat mesquit will be knee high and well matured.[67] I wish you had the mowing of it this fall, for I do not need it—shall not touch it.

As yet I have had no news of the arrival of the bark Texana, with the bucks, and as they had a terrible gale on the Florida coast about the 10th inst. and as she has been out over a month, I begin to feel quite uneasy about her safety. And even if the vessel reaches port, there is much danger that the bucks, (probably on deck,) may have been washed overboard. I will write you immediately on receiving news, whether good or bad.

I enclose a draft for $75 on Westray, Gibbes & Hardcastle, of New York, in favor of Gen. O. F. Marshal [68] which please send to him in payment for the Honest Ram, and say to Gen. M. that I will write him a letter on the arrival of the buck, should he fortunately reach me. He would find glorious picking were he here now—such as he has seldom had in Steuben county.

It is unfortunate that the bucks were not shipped on the schooner Stampede, which sailed from New York six days *after* the Texana. Striking favorable winds, she got in some ten or twelve days ago, or just before the gale.

---

[67] "In the western portion of the State, the long mesquite makes a hay that can hardly be surpassed; and all that is required to produce this excellent grass for hay, in the greatest abundance, is to fence in the land to protect it from the stock running at large."—William J. Jones, "Grasses for Hay, Corn Fodder, Etc.," in *The Texas Almanac for 1869*, 135–36.

[68] Otto F. Marshal (1791–1891), a farmer (his title came from service with the state militia) of Wheeler, Steuben County, New York, from whom Randall obtained "Old Honest" for Kendall. Marshal was for years a friend and correspondent of Randall, sharing with the latter an intense interest in and devotion to sheep. He wrote to Randall 1 June 1860: "As we have not sheared Honest I concluded to send him in his wool [.] I should be pleased to have him go to Texas in his fleece if he went safe [;] he would make such a grand appearance in his fleece if not pulled out to much on the rout [.]" Randall Papers. Although Marshal wrote to Kendall about "Old Honest" and the sheep of southwestern New York, he apparently never had a reply.

I shall start up to my sheep estancia the moment it clears up—I must be there on the 1st September, to wean the balance of my lambs and deliver some 500 old ewes which I have sold.

Truly yours,
Geo. Wilkins Kendall

*Aug. 28.*–We have a clear and bright day at last. Providence be praised: for a week the sun has hardly shown his face, and the rain has prevented me from going down town. Never have I seen corn growing so fast as now: I am planting the early Canada, and will let you know whether I make a crop or not.

I will not seal this letter until I get down town: if I find a letter at the post office giving any news of the bucks I will add a line.

Yours ever
G.W.K.


Rancho near New Braunfels,
Sunday, Sept. 9, ½ past 3, P.M.

My Dear Sir:–Fifteen minutes ago the mule wagon I sent down to Indianola after the bucks came into my yard, in five minutes more *all* the animals were once more on Mother Earth, and in less than another minute two couples of the younger paired off and set to fighting like mad, the Old Honest Ram, albeit clumsy and unsteady on his pins, chassering and dos-a-dosing about as if inclined to do battle with any or all. I presume it is almost the only time one of them has touched terra firma since leaving Cortland Village, each has undoubtedly gone through great personal inconvenience on the way, each very probably looks upon his neighbor as the cause, and hence their pugnacious set-tos on reaching the ground. They came up in less than five days from Indianola, and, although somewhat "gaunted," still appear to be healthy and in fair condition. Well, I am glad they are all here, after the rain storms and tempests we have had, and knowing that you also will be rejoiced, I hasten to communicate the fact.

*5 o'clock, P.M.*–After browsing about for an hour in the yard, and after having settled, for the time, their private disputes, the bucks are all lying down quietly under the shade of a spreading live oak in my yard, and are chawing their cuds as cosily as possible. Two or three of the yearlings seem a little shy—at least I cannot get near them. This will wear off in a day or two, with gentle treatment, and I shall probably keep them here at my rancho for a month, giving them as much grass as they can eat, and a little corn meal and water every day.

I came down from my sheep Estancia in Blanco county yesterday: left every thing, as usual, in tip-top condition— all the sheep, old and young, as fat as seals. You are aware that I have sold down to no other than young, active, and thrifty animals—all my old ewes and wethers have been disposed of, and I shall enter the coming winter with over 2000 breeding ewes as fine, all things considered, as you ever laid eyes upon. My last spring's lambs are all weaned, (about 900 bucks and wethers in one flock and as many ewes in the other,) and the latter will not see a buck until the fall of '61. There are beauties among them: many look like grown sheep, although barely five months old. For sheep, at least near my place, we have had a glorious summer—never any thing like it. There *is* balm in Texas.

*Sept. 10.*–The bucks are now running, or hobbling about the yard; they have not as yet "got their land legs on" after being so long cramped, cribbed, cabined and confined on ship-board. Why should they be "mustangy" as we say here? or wild and shy: I can hardly get near them, and the "Honest Ram" is the wildest in the lot. My wildest sheep are not more shy: it may be that they have been annoyed and pestered on the way, or else that they do not understand liberty: perhaps they have some lurking presentiment of evil, or else think they are lost in a wilderness of prairies. I shall endeavor to gentle them as soon as possible, and wont them to their new home.

I have not as yet had time to examine the yearlings; I shall not pass judgment on them at present, as it would be unfair. No. 6 is in altogether the best condition—is the broadest and most compact—and seems to me to be the best buck in the lot: he has an independent, saucy, devil-me-care way, and fights manfully for his meal. No. 384 comes in the poorest condition—he is very thin in flesh, and has a gangling, weakly look. He may mend in appearance, and show more strength, with more flesh. No. 117 comes next to No. 6 in condition; the others are all in fair order—much better than could be expected after a journey of some 2500 or 3000 miles, and of nearly two months duration. In the course of a week or ten days I will write you more fully as to their appearance and condition.

Your long and interesting letter of 20th ult. came to hand just as I was leaving on my last trip up. In it you ask "Are considerable numbers of men being *hung* in Texas as abolition conspirators and incendiaries? Are there such conspiracies and incendiarism?" &c, &c. True, every word of it. Northern jack-leg Methodist preachers have been hung by scores: they were caught in the act of giving the negroes in Eastern and Northeastern Texas poison and pistols, and inciting them to murder, rape, arson, and other atrocious acts, *and they were hung* as fast as caught. For them there is no pity: but for the poor negroes, their dupes, who have been hung, there is sympathy: they were happy and contented until tampered with by the accursed spawn sent out by the Northern fanatics.[69]

---

[69] Many Texans were in the throes of intense excitement approaching hysteria during the summer of 1860 as a result of rumors of abolitionist activities. Fires attributed to abolitionists wrought damage in a number of towns. Stories of insurrections and poisonings by slaves

Here in Western Texas we have had no trouble save with
Mexican and other horse thieves, and they all come to the
same end as the Methodist preachers in the East—a ropes
[sic]. Lynch law, in a new country, is better than any law
formed by the wisdom of man since the Mosaic—has fewer
flaws and quibbles, and metes out more even justice.

Yours    Kendall

My every moment of time is taxed—I will soon endeavor
to send you a long letter.

I shall make a second crop of corn if the bud worms do
not destroy it.

Rancho near New Braunfels,<br>
Monday evn'g, Sept. 10, '60.

My Dear Sir:–I placed a letter for you in the post office
this afternoon, announcing the safe arrival of the bucks—
all of them: I at the same time took out yours of the 27th ult.,
describing your trip down from Syracuse in an afternoon
train, during a "smart chance" of a shower. The morning
of that very day, as I see by looking over my note book,
found me making a fine pot of English breakfast tea, (I *will*

---

circulated. Vigilance committees were organized, and some lynchings,
both of Negroes and whites, occurred; the total number would appear to have been small. Some Northerners believed that the rumors
resulted from a deliberate effort to quiet anti-slavery sentiment in
Texas; some Southerners saw in them an effort to weaken Unionist
sentiment in Texas and thus undermine support of the Bell-Everett
ticket in the impending election. See Allan Nevins, *The Emergence of
Lincoln* (New York, 1950), II, 307–8; Claude Elliot, "Abolition in
Texas," in *The Handbook of Texas,* (Austin, 1952), I, 3.

Two days after Kendall wrote this letter, John Adriance wrote to
Randall from Columbia, Brazoria County, Texas:

"The diabolical plots of Abolition emissaries from the North have
left sad and unmistakable traces of their hellish work upon our pros-
trate frontier.

"Rumors may have been unfounded, and a few facts exaggerated,
but in the aggregate, the *worst* has not been told."

have good tea,) miles from any house, camped near a spring, and before sunrise I was harnessed up and on my way to my sheep estancia. And a glorious day it was, the prairies and gently sloping hill sides, clothed in richest green, and the grass growing with a luxuriance of which you poor Northern folk have no idea.

You must have had a time of it during that ride—your description of the "peltings of the pitiless storm," [70] is graphic to a degree. While reading it, a confused run of ideas got in my head, and I thought of Mazeppa's ride,[71] and Tam O'Shanter's,[72] and scenes from the Pilgrim's Progress, and lines from the "Star Spangled Banner":

> "The red rockets' glare, [sic]
> Bombs bursting in air,"

and various games of that sort, as the noted Samson Brass [73] would say. I have been out in just such showers, many a time and oft, and had to take them. The longest shower I was ever exposed to was in June, '46, while campaigning in Mexico. I was then a "high private" in Ben. McCulloch's company of Texas Rangers,[74] and after the occupation of Matamoros Gen. Taylor,[75] anxious to learn how many troops Gen. Arista had at Linares, some 150 miles inland, sent Mc-

[70] Shakespeare, *King Lear*, Act III, Sc. 4.

[71] Mazeppa, the hero of Lord Byron's poem *Mazeppa* (1819), was lashed to a wild steed, and freed only after the horse died at the end of a mad run.

[72] Tam O'Shanter, in Robert Burns's poem *Tam O'Shanter* (1791), rode furiously to escape witches which were pursuing him; he was safe after crossing the middle of the river Doon.

[73] A shyster lawyer in Charles Dickens' *Old Curiosity Shop* (1841).

[74] Ben McCulloch (1811–1862) was well known in Texas as an Indian fighter before the Mexican War. During the war he organized and led a company of horsemen which performed valuable services. Kendall's despatches to the New Orleans *Picayune* helped greatly to give McCulloch a national reputation.

[75] General Zachary Taylor (1784–1850) was encamped near the Rio Grande when the war with Mexico broke out. In May, 1846, he occupied Matamoros, a Mexican town on the other side of the river.

Culloch to ascertain. There were 35 of us in all, all well armed, and every man had a horse under him on which "you could run for your life," and save it. We were fourteen days on the scout, and it poured showers all the time—with intermissions. But these intermissions were so short that we hardly had time to dry our clothes—the showers run into each other, like the confluent smallpox. I slept one night, with my saddle for a pillow, and was wakened only by the water running into my mouth: the shower came down in a chunk, like a water spout. On the morning of the 15th day we entered Reynosa, then occupied by our troops, as hard a looking set of customers as the rear guard of Peter the Hermit's crusading host: I had not taken off boots, jacket nor cravat for a fortnight, had not been under shelter, and my only regret was, that I could not have my daguerreotype taken just as I was.

This was the longest spell in the way of a shower I was ever caught in. In '37 I chartered a pilot boat at Charleston to take a party of eight of us to Norfolk. Off Hatteras, a terrible gale struck us; I am not troubled with that peculiar idiosyncracy [sic] of the mind which superinduces sea-sickness: I could not stand the foul atmosphere of the little 7-by-9 cabin; and so I lashed myself to the main mast and trunk of the vessel, and under the lee of the aforesaid trunk I sat for seven long days and nights, drifting down the Gulf stream, ruminating, meanwhile, on the vicissitudes of human life, on the duration of storms, and the speed of ocean currents, and eating hard biscuit and raw onions for the want of any thing else. But what do you care for all this? I will stop. Good night.

*Sept. 11.*–The bucks are picking up smartly: old "Honest" peers at me with a sinister look whenever I go near him, but has as yet made no hostile demonstration. He may "run agin" me one of these days, and lay himself amenable to an arrest or charge for assault and battery. And his battery I can dispense with. As for his shape I can say nothing: with his great coat on he looks like the butt end of a saw log, cut off some 3½ or 4 feet long, and with four huge pegs drove in on one side. Yet I like him hugely. Of the younger animals I still affect No. 6, the one with the tuft of wool left on the shoulder. He is in tip-top condition, and strong as a moose.

I am now ploughing and hoeing my second crop of corn, which is nearly shoulder high and on the point of tasseling. The bud worms are at work upon it: barring their ravages, I am almost sure of a good crop. The weather is glorious for corn, and you can hear it grow of a still night.

*Sept. 13.*–My head shepherd, Tait, called last evening, and remained all night. He thinks the world of "Old Honest," and also of No. 6, or Capsicum as I intend to call him. (Is not Capsicum one of the main ingredients in Dr. Thompson's No. 6?) [76] I don't like the name of Lobelia. No. 384 is the only buck in the lot in right poor condition, and I am a little fearful he was injured in some way on the route. He shall have all attention.

As a matter of policy, I do not deem it best to say that any of the younger bucks are for sale at present: they do not show to any thing like the advantage they will hereafter, and will hardly be in condition to serve ewes this fall. I seldom use a buck until he is 2½ years old. It will cost me comparatively nothing to keep them all until another year, and in the best possible condition. What do you think of it?

I have scratched this off in great haste, by snatches.

Ever yours
Geo. Wilkins Kendall

---

[76] Samuel Thomson (1769–1843), a New England botanic physician, was widely known for his method of treating human ailments. Believing that cold produced disease, he sought to increase the inner heat of the patient's body. The herbs *lobelia inflata,* which induced vomiting and perspiration, and *capsicum* (Cayenne pepper) were favorite remedies. He patented his process and licensed other practitioners. His writings include *New Guide to Health: or Botanic Family Physician* (1822). See *Dictionary of American Biography,* XVIII (1936).

Rancho near New Braunfels,
Sept. 28, 1860.

My Dear Sir:–Talk about the sprinkle you had that night
you rode home in the rail car! Why, it was "cakes and gingerbread" compared with a shower I faced yesterday, in an
open wagon, on my way down from my sheep estancia. I
have quoted a remark made by my friend Sancho Panza:
do you recollect it? It was the morning after Sancho had
received the terrible pummeling at the inn that his master
asked him how he had passed the night? "Why," quoth the
Squire, "the pack-stave threshing I received yesterday was
cakes and gingerbread compared with what I received last
night." It is years since I have read "Don Quixote," and I
may not have quoted the exact remarks of the redoubtable
Sancho; but I have given the substance: the chapter is one
of the most laughable in the whole book. But to the storm
of yesterday.

I have a favorite mare, Bonita—game all the way through.
She is afraid of nothing—would face Garibaldi [77] or the
devil—I could drive her straight up a lightning-rod or liberty
pole if she had claws, even were there a church on fire at
top. But yesterday, as we were facing the shower, poor Bonita
said it was too much, and turned square round in the prairie, and there stopped. I turtled. I had on nothing but a blue
linen blouse and shirt of the same—not much protection
with the rain coming down at an angle of about 45 degrees
and in a solid chunk. I say I turtled, i.e., I drew my head as
far in as I could, to keep the water out of my neck, and sat
there in the wagon until the storm was over, drenched as
wet as a drowned rat. As the shower eased I took a "smile"

[77] Giuseppe Garibaldi (1807–1882), the Italian patriot, was making
history during the spring and summer of 1860 by his conquest of
Sicily and Naples. Three weeks before Kendall wrote this letter,
Garibaldi had made his entrance into Naples accompanied by the king
of Sardinia.

of old Bourbon, lit my pipe, (not without difficulty,) and whistled along home. Dry clothes, a cup of black tea, and a plate of delicious mutton chops completely "resurrected" me, to use a common but not very classical expression, and a good night's sleep brought me out as chipper as a cat-bird. There, I'll put my shower against yours, and bet that I beat you. I was not cooped up in a close rail-road car; I was out where I could see, hear, feel, and enjoy (!) all. Selah.

What gorgeous, glorious weather we are having for grass— never have I seen it so luxuriant, even in Texas. My second planting of corn is now all silked out, and ears forming: were it not for the bud worms, which are at work upon it, I should make another heavy crop. We now have an abundance of butter and snap beans, sweet potatoes, okra, tomatoes, and green peas, with other vegetables, and such melons as you never tasted North of Mason and Dixon's. Why, I have musk-melons (French) in my field, "laying about loose," which you could not get into a bucket, and water melons which would fill a 10 gallon keg. There *is* balm in Texas—occasionally.

*Sept.* 29.–I have taken a good look at all the bucks you sent me this morning, and find all thriving wonderfully well —all getting fat except No. 384. He feeds well enough, and holds his own, but does not improve so fast as the others: I am confident he has not the stamina. No. 6 is altogether the heaviest of the lot, and is a hearty and industrious feeder. Nos. 117 and 144, the wrinkliest of all, are in excellent condition, and very promising. Now about 364, described by you as shearing 9 lbs., as half brother to 6, or "Capsicum," and as being a peculiarly short-legged, hardy fellow. *No such buck has reached me.* The numbers on four are plain enough: 6, 117, 144 and 384. The number on the fifth is plainer than on any of the rest—183 (One Hundred and Eighty-Three.) How is this? If No. 364 started, he was certainly changed on the route. The man who brought them up for me, and who took from the vessel such as were delivered to him as for me, says there were other bucks on board. Has there been cheating? Did some one on board the bark Texana fall in love with No. 364, and surreptitiously put No. 183 in his place? If he was on the bark this must have been the case, unless the animals got loose or mixed together while on the passage, and by mistake No. 183 was placed among my lot.

Now, what am I to do? No. 183 is a thrifty buck enough, but no more to be compared to No. 6 than I to Hercules or Gen. Jackson, while you describe the missing No. 364 as the largest animal in the lot, and the heaviest shearer. If there has been "rough gambling" while the bucks were on the way I wish to trace it out. Unless the figures 364 have been obliterated or defaced, the buck in question is now carrying them and can be traced, especially if in Western Texas, and I will expend his value in the attempt to find him.

I regret that I did not ascertain this sooner, and write you at once. Expecting an answer by mail, I subscribe,

Truly your friend
Geo. Wilkins Kendall

The "Honest Ram" is in good plight—knocked my oldest son, a lad of eight years, heels over head yesterday. Cause unknown.

I am busy as a man well can be—am building several houses and shelters, and my every moment occupied.


Rancho near New Braunfels,
Oct. 1, 1860.

My Dear Sir:—I see by the last Galveston News that nine bucks, from your flocks, were lost the other day at Richmond in this State: I have not the paper now by me, but the paragraph states to the effect that [sic] "that 9 Merino bucks, from the flocks of Hon. Henry S. Randall, of N. Y., and belonging to Mr. Blassingame [78] of this State, were drowned by the falling in of the Railroad bridge at Richmond.[79] Six others died on the passage out, or were lost in some way." Such was the substance of the item in the News. A family by the name of Blassingame formerly lived in the upper part of Blanco county, but I believe they have moved to Caldwell county, not far from Lockhart.

I am still utterly at a loss to account for the non-appearance of No. 364, and for the appearance of No. 183 in his place, and shall continue to be uneasy until I hear from you. I presume that the captain of the bark Texana, described to me as a worthy and reliable man, is now either in New York or on the way there, and he might give some clue to unravel the mystery. For whom were the other bucks on the Texana? and who had charge of them? I have heard of children being changed while at nurse, of children being carried of [f] [by] old Gipsey [sic] women, and all that sort of thing, but never before have I heard of bucks being changed while en route. Could No. 364 have jumped out of his quarters,

---

[78] John C. Blassingame, of Moulton, had expected Randall to ship him some sheep in the spring, but Randall had not sent them until mid-summer. There is considerable correspondence about Blassingame's sheep in the Randall Papers. Blassingame's sister Aurelia was married to Benjamin Fitzpatrick (1802–1869), who in 1860 was a United States Senator from Alabama. Blassingame had written to Randall after reading the latter's article in the *Texas Almanac for 1860*.

[79] A town in Fort Bend County, on the Brazos River, southwest of Houston. By the end of 1860 the Buffalo Bayou, Brazos and Colorado Railway, the first railroad in Texas, had been completed between Harrisburg, near Houston, and Alleyton, near the Colorado River, a distance of eighty miles. A temporary drawbridge only a few feet above the water had been built across the Brazos at Richmond. Since the bank on each side was about thirty feet high, the train approached the bridge on a steep inclined plane, gathering momentum on its descent for its upward climb. On 21 Sept. 1860, something gave way, and the locomotive and several cars were thrown into the water or against the bank. See the *Houston Tri-Weekly Telegraph*, 22 Sept. 1860; S. G. Reed, *A History of the Texas Railroads* (Houston, c.1941), 53–65.

and No. 183 have jumped into them? I cannot get it out of my
head that there has been "rough gambling," or what Capt.
Simon Suggs [80] would call "shykeenery," somewhere, and
am most anxious to ferret it out. Put me on the track if you
can.

I shall start up to my sheep estancia to-morrow or next
day, taking all the bucks with me. The "Honest Ram" I shall
put immediately to work, marking all the ewes as fast as
served: it is early, but I wish to get as many lambs from him
as possible before 15th November, when my season will be
over. At first I shall give him the very choicest of my Merino
ewes—(I have about 150 full bloods)—and these I shall
shelter and watch next March.

October broke in upon us this morning bright, balmy and
beautiful: during the day showers have been passing about,
and I am fearful, (yes, fearful!) that more rain is to come.
Think of a Texan afraid of rain!

Do you get the Weekly Picayune? I asked one of my partners, in a recent note, to send it to you. The reason I ask is,
that I have recently written two or three letters for the Pic.
in answer to certain croakers and pretended "victims" of
mine, which I am anxious you should read.[81] I scratched

[80] A fictional character created by Johnson Jones Hooper (1815–
1862). The first story about Suggs, a shrewd, shifty backwoodsman,
appeared in 1844, in the *East Alabamian,* a weekly edited by Hooper
in Lafayette, Alabama. *Some Adventures of Captain Simon Suggs,
Late of the Tallapoosa Volunteers; Together with "Taking the Census,"
and Other Alabama Sketches* was first published in 1845. See W.
Stanley Hoole, *Alias Simon Suggs, the Life and Times of Johnson
Jones Hooper* (University, Alabama, 1952).

[81] There had apparently been complaints that Kendall had misrepre-
sented Texas, and by painting too strong a picture of her resources
and possibilities, had made himself responsible for the ruin of some
who had taken him at his word. Letters by Kendall in answer to the
"croakers" appeared in the *Daily Picayune* on 23 Sept. and 7 Oct.
1860. In the first he declared: "I have never described Western Texas
as a paradise on earth, or anything of the kind; but after having visited
every State and Territory of the Union, save California, Oregon and

them off hastily, but made some rough hits, to my thinking, at certain shiftless, worthless shoats, who have been decrying Texas, and who will run down any region where they cannot obtain a fat living without work.

Why, my dear sir, if it were in the power of man to guarantee me precisely five such years ahead as the last five have been: the same amount of heat and cold, of rain and shine, of grass and water: if, I say, I could have secured to me precisely such years until 1865, and have my health and strength meanwhile, I would cheerfully give bond and security to pay to any orphan asylum the sum of $20,000 per annum. How afford it? you may ask. Why, I would cover every hill and valley in four counties with sheep, and at the end of five years, in case I wished to leave Texas, my only trouble would be to find transportation for my money.

What a sad calamity was that on Lake Michigan! [82] And

---

Washington, I believe that we are nearer paradise than the people of any other portion of the United States, and I can never be sufficiently thankful that a kind Providence bent my steps hither." He further asserted that for the four years ending in May, 1860, he had realized a clear profit of 75 per cent per year on his investment. In his second letter he told of trips which he had made in the 1820's and later through Ohio and other parts of the West, emphasizing the frontier conditions then prevailing in those parts; Texas was now going through a similar period.

He had earlier (*Daily Picayune* 27 July 1858) written scornfully of some of those seeking new homes in Texas: "Many wish to find good macadamized roads, churches of their own denomination, colleges, schools, the society of an old settled community, *and good land adjoining at one dollar per acre.*" Such people, he advised, should stay home or go to Kansas.

[82] The sinking of the *Lady Elgin* on Lake Michigan, 8 Sept. 1860, twenty minutes after being struck by the schooner *Augusta*. Nearly four hundred people lost their lives, Francis Lumsden, his wife, their son, an adopted daughter, and a servant, among them. This was the second greatest ship disaster in Great Lakes history. Lumsden's body was recovered more than a month after his death, and a great funeral held in New Orleans.

A moving letter from Kendall about the death of the Lumsdens appeared in the *Picayune* 9 Oct. 1860. Lumsden, a North Carolinian, had

what a cruel loss did I sustain in the death of my old partner, Lumsden, who went down with his entire household. As I read the accounts, I felt as though I could set down on the door step, and cry. We started the Picayune together, did Lumsden and myself, nearly a quarter of a century gone by, and were ever fast friends.

I will write you again on my return.

> Truly yours,
> Geo. Wilkins Kendall

P.S. Just as I was closing this, the man who brought up the bucks passed. I called him in, and questioned him. He said that at Indianola the captain of the Texana told him that he had brought out 31 bucks all safe, 25 of which were taken off at Saluria at the entrance of Matagorda Bay. So, they must have gone either towards Corpus Christi, or the Nueces, or else on the peninsula.[83] I shall make no move until I hear from you.

> In haste,
> Yours &c
> G.W.K.

---

served his newspaper apprenticeship on the Raleigh *Register*. During the winter of 1833–1834 he and Kendall were both working on the *National Intelligencer* in Washington, and there became strong friends. They met again in New Orleans, and finally in 1837 founded the *Picayune*. In 1860 Lumsden was no longer active in the management of the paper; he traveled a good deal, and wrote some letters for the paper. In addition to Kendall's letter see the *Daily Picayune* 14 Sept. 1860; Dabney, *One Hundred Great Years, the Times-Picayune from its Founding to 1940;* Copeland, *Kendall of the Picayune.*

[83] Saluria, at the northern end of Matagorda Island, was another of the small towns which aspired to be the leading port on Matagorda Bay. It was burned by the Confederates during the Civil War and never rebuilt. Francis Lumsden described it (*Daily Picayune* 15 Feb. 1860) as "the little (very little) town of Saluria, situated upon a point of a large (very large) prairie. . . ."

Corpus Christi is on Corpus Christi Bay, at the mouth of the Nueces River.

The peninsula referred to is that which separates Matagorda Bay from the Gulf of Mexico.

Rancho near New Braunfels,
Oct. 6, 1860.

Talk about a ram at a gatepost! my dear sir; why, "Old Honest" is at least 40 ram power, and could have carried away the gates of Gaza, and Samson along with them, had he been on the spot. Mad! he was madder than 40 March hares three evenings ago, when I arrived safely with him at my Estancia, as he saw the different flocks wending their way into the different folds. He saw a deal of work before him, and although just off a rough mountain journey seemed raving anxious to make a commencement. The night before I "camped out" on the way, spreading my blankets on the ground. During the early part of the evening a brisk shower drove me under the wagon for shelter, and during the rest of the night he was pawing over head (for I did not take him out of the wagon,) like a wild Mustang stud. Were I not constitutionally opposed to a change of name, whether in a human or a brute, I should certainly alter his; I should call him either Samson or Il Furioso!

Speaking of names, puts me in mind of an anecdote. One hot forenoon in 1846, while Gen. Taylor was at that hottest of all holes Camargo, preparing for his march on Monterey,[84] a wag of a devil belonging to the Commissary's Department asked me if I had half an hour's leisure, as he wished my advice on a point which he could not well decide. I declared myself at his disposal; we walked out about half a mile under a broiling sun; he took me up to a little swinge-tail, calico, Mexican pony, with blue-and-white crockery eyes—worth about seven dollars and a half—tied in the chaparral outside the town, and asked me to examine him! said he wanted my opinion upon a point of much importance! "D-n the pony," said I, "he isn't worth an opinion." "Don't be hasty," retorted the wag; "what I wish you to decide for me is, which of two

---

[84] General Taylor established a base at Camargo at the mouth of the San Juan River in July 1846; in September after a hard battle he captured Monterey.

names I shall call him: I was thinking it over all last night after I had bought him, and finally resolved that henceforth and forever he should bear one of two names, and you are to decide: *shall I call him Spriggs or Sesostris?*" [85] "Sesostris!" retorted I, "if he can pack it," and this ended the colloquy. Of course I was not fool enough to get mad at the wag's impudence, and have since laughed a thousand times at the sequel. Pardon this episode, and–but, while my hand is in, let me tell you another anecdote of an impudent acquaintance of mine.

On the Santa Fé Expedition, in '41–2, we had a character by the name of Tweed, Jimmy Tweed. Jimmy was born at Gibraltar—his mother an Irish washerwoman and his father probably an English officer: at all events, Jimmy had a fair education, and wrote a beautiful hand. He served out one or two enlistments in the British army, came over and served one enlistment in our service, and was all through Florida. The next turn of the cards found him in Texas, and enlisted for the expedition to Santa Fé, and being an excellent soldier and penman was made sergeant-major. Jimmy was a genius: he possessed an inexhaustible fund of humor, a great deal of shrewdness, was a thorough judge of human nature, and as thoroughly unprincipled as a man can get to be in this corrupt world. After we were taken prisoners, and on the long and dreary march to the City of Mexico, Jimmy "froze" to me—he knew that I had money, wanted his daily drinks of Catalan brandy, and could always continue to draw on my pockets for any little sums he needed—there was no resisting his importunities. We were all finally released, and I lost sight of Jimmy for two years, when I received the following laconic but extremely plain and pointed epistle from the scamp:—

---

[85] Sesostris was a legendary king of Egypt who was credited with great exploits. Kendall himself had a horse named Spriggs while he was in Mexico.

"Houston, Texas, Aug. 1844.

"My Dear Kendall: I have married an old faggot here with
"a little money: so soon as that is gone, of course I shall
"leave. What are the chances for a single man of my personal
"appearance and education in New Orleans?

"As ever faithfully yours,
James Tweed."

There, if you can find, either in ancient or modern history,
a piece of more cool, unblushing, atrocious impudence than
that, I should like to know it. Admitting his marriage, announcing his intention to desert his wife so soon as her money
was spent, and then asking me what the chances were for
a *single*(!) man in New Orleans! I could not answer his letter
well, and have never heard from Jimmy since. Think of *his*
addressing "My Dear Kendall."

Good night. I will add a page or two to-morrow—may
perhaps give you something more interesting.

*Oct. 7.*–When I left my estancia, early on the morning of
the 5th inst. I ordered my head man to pick out all the pure
blood Merinoes at once, and set "Old Honest" immediately
to work *a la stallion.* I am going up again to-morrow to see
how he is getting along. I have not sheared him, and it was
amusing to see his performances when I first turned him
loose in a pen attached to my large shed. In an adjoining
pen were some half a dozen crippled ewes, and to get at
them the old fellow went butt against the fence, starting
one of the boards at the first dash. I then caught and tied
him to a post, which he pulled down at a single jerk. I next
told one of the shepherds to bring a couple of buck lambs,
nearly half grown, to keep him company. The moment they
set eyes on him they stampeded, ran, slunk away in a corner,
and seemed half dead with fright! they evidently took "Old
Honest" for a bear. And well they might, for, as you said in
a former letter, he looks like one.

I am going up again to-morrow morning, and on my return will inform you how the old fellow is getting along. I gave orders to have all the ewes marked as fast as served, so that I can pick them out by the 4th of next March, and by the 1st November I am in hopes all my full bloods will be with lamb.

The young bucks remain here at my rancho, and are not very popular with my wife and the other women folks about the house. When I inform you that they have stripped all the young peach trees as high as they can reach, and all the rose bushes in the yard as well, you may possibly divine the cause. They are all, save No. 384, as fat as seals, and growing wonderfully. I shall send them all up to the estancia in the course of a few days. I am now cutting hay—fine, black mesquit—and it will be better than ever. I could mow 5000 acres, grass nearly knee high, if I needed it: as it is, I shall cut some 8 or 10 tons of choice.

Our weather continues warm, growing, glorious—never experienced any thing like it. My second crop of corn—in the ground not quite two months—is now in roasting ear: I should have had a tremendous yield had it not been for the bud worms.

My tomato vines are fairly breaking down under the load of a second crop; melons are rotting in the field; of all sorts of vegetables we have an abundance—okra, squashes, snap and butter beans, radishes, cucumbers, &c. &c. By the papers I see that some two weeks ago you had frost in New York, ice half an inch thick in some places, and snow on the mountains. God forgive the people who live in such a climate—they know not what they do.

I will write you again soon, and meanwhile believe me,

Truly your friend,
Geo. Wilkins Kendall

Had I put in a field of spring oats or barley on 7th August, I should have made a magnificent crop.

Rancho near New Braunfels,
Oct. 11, 1860.

My Dear Sir: Hyena, after all, would be a better name
than either Il Furioso or Samson for the "Honest Ram." You
have probably seen one of those mild-featured, amiable
African brutes promenading around his cage in the menag-
erie: so of old "Honest," for he is eternally and everlastingly
walking round in a circle as far as his rope will reach, furnish-
ing as good an example of perpetual motion as you could
well imagine. If we let him loose, he is instantly at work at
the fence: if we tie him, he winds himself up, and has already injured one of his eyes so badly that I am fearful he
will lose it. He has already knocked the man who attends
to him heels over head twice, and is constantly performing
other eccentricities. We give him eight or ten ewes a day;
but like children after Sherman's lozenges he is ever crying
for more. After a few days, I shall turn him into the flock,
(about 160,) and let him help himself. I only trust that he
will not get a set of lambs as fretful, restless and uneasy as
himself: I like "Old Poll's["] manners much the best.

I left my sheep estancia yesterday—went up with Mrs. K
and one of my boys. My wife was anxious to see how our
new houses look, and so I took her up: we "camped out" two
nights, and came back none the worse for wear. I think you
could stand it for a night or two, and live through. In the
course of three or four weeks I am in hopes we shall move
up, and then farewell to the old "Rancho near New Braunfels" and its pleasant surroundings. I wish I could send you
a basket of roasting ears and tomatoes, both of the second
crop: we are now reveling on the good things of this life.
Melons! Mon Dieu! My second crop of corn would have been
very heavy had it not been for the bud worms. I am now
mowing hay, and putting in a field of barley for winter grazing, anticipating a good yield in the spring. *Nos veremos*, as
the Spaniards say.

All the young bucks, save No. 384, are fattening and thriving: Capsicum, (No. 6,) the fattest and heaviest of the lot. Even 384 is improving, and he may come out this winter. But I doubt whether he will ever turn out as likely as the others. In a day or two I shall start up to Post Oak Spring with them.

I have learned that all or a portion of the 25 bucks left by the bark Texana at Saluria were for a Mr. McCreary [sic],[86] or some such name, who has a flock of sheep on St. Joseph's island.[87] Has he the missing animal, think you? I shall not make a movement until I hear from you.

I have all along been in hopes that I should be able to pay

---

[86] A James K. McCrearey received a grant of land in Austin County in 1839. In 1841 he was living in San Felipe, and was clerk of the district court; in the obituary of his wife, who died in that year, he was referred to as Dr. McCrearey. He was elected Chief Justice of Austin in 1842 and held the office until his resignation in 1844. He was a member of the legislature in 1844–1845. He became Commissioner of Pilots for the port of Paso Cavallo in 1849, holding this position until 1852. Here the trail ended for the compilers of the *Biographical Directory of the Texan Convention, 1832–1845* (Austin, 1941).

The Dr. James K. McCrearey concerned in the episode related in the Kendall letters lived near Saluria on Matagorda Island, at the entrance to Matagorda Bay. At the beginning of 1860 he had a flock of some 500 sheep, Mexican ewes crossed with merino and Southdown rams. Having read and liked Randall's article, "Sheep Husbandry in Texas," in the *Texas Almanac for 1860*, he wrote to Randall expressing his desire to buy five or six rams and ten ewes. At this time he was planning to leave the coast in the fall and establish himself in the hills above Austin. Randall offered to sell him five rams and twenty ewes for $300, five rams and ten ewes for $210, 100 sheep for $800, 50 for $500, 25 for $300. McCrearey ordered twenty ewes and five rams; he sent Randall $50 and an order on a New York commission house to turn over to Randall the proceeds (or part of them) of the sale of 1,600 pounds of wool belonging to McCrearey. After a delay occasioned in part by a sluggish wool market, Randall sent McCrearey twenty ewes and five rams on the bark *Texana* which sailed from New York about 20 July and arrived off Indianola 36 days later, carrying also the sheep of Kendall. Correspondence between McCrearey and Randall, Randall and the New York commission merchants, and other letters bearing on McCrearey's sheep are included in the Randall Papers.

[87] A large island down the coast from Matagorda Island.

your section of the country a flying visit this fall; but the delays ever incidental upon building, and other matters, will now probably prevent me. But if I live, you will certainly see me next summer, after shearing time is over, when I shall have more leisure. I hate a crowd, and shall probably reach New Orleans after the spring travel is well over.

We had a little touch of a Norther day before yesterday, bringing along a few stray flocks of wild geese along with it: to-day the weather is warm and growing again, with some signs of rain.

I will write you whenever any thing of interest turns up. In haste,

Truly yours,<br>Geo. Wilkins Kendall

Rancho near New Braunfels,<br>Oct. 19, 1860.

My Dear Sir:—Well, the Hon. Samson Hyena Il Furioso, Esq. has mastered all hands at last, and is now having every thing his own way spite of bars, bolts, ropes, &c. &c. Three nights ago I reached my estancia, when my head man, an honest German, said the old ram had twisted and broken all the rope on the place, had knocked him over again and again, and that he could do nothing with him. I went into the shed, just at dark, and found him tied with a new piece of rope which would hold a Bull of Bashan. The next morning, on going out to the pen, he was missing, a piece of the rope only left. On looking at the fence separating him from the ewes, I saw it was knocked flat: on looking farther, I saw another gap, through which "Old Honest" had made his way, taking all his ewes with him! Here was a job. After a hunt in the prairies, we finally found all, and all safe: but a straggling wolf or panther, or a stray dog, might have had glorious picking had such "varmints" been out at the time. Since then, I have allowed the old fellow to run regularly with the Merino ewes, and I am in hopes he will serve the most of them before the season is over. As a well behaved, manageable buck, "Old Poll" is worth a dozen of him. I turned him in, year before last, with 138 ewes, and in three weeks time he got them all with lamb, living on grass alone and not losing an ounce of flesh. He has his conceits, has "Old Poll," and among others will have nothing to say to a ewe a second time after having once told his story.

The five young bucks are still at my rancho here, but I shall start up with them to-morrow or next day. All, save No. 384, are as fat as seals, and even he is mending. They have stripped every peach tree as high as they could reach in our yard, and about the house, and every rose bush as well, and are not at all popular with the women folks.

I have not offered any of them for sale: nearly all my neighbors about here are supplied for this season, and when the animals first arrived they were in no condition to show. Had they reached me a month earlier, it would have been a different matter. What are the four worth? Will $100 pay you for them? If so, I will send you the money at once. They have cost me, to bring them out, over $20 a head, and I doubt whether I could obtain over $50 a head for them next fall—few about here are willing to pay even that price. One reason is, that there are few good judges of sheep among them—few who have had experience.

I am heartily rejoiced to learn that there is some chance of a fusion in your State, and as I see that you are nominated as an elector at large most sincerely do I hope that you may be elected.[88] There is one man in your State I wish to see

[88] Randall was one of the members of the New York delegation to the Democratic National Convention who returned dissatisfied with the results. Although he had favored the Wilmot Proviso, been a Free Soiler in 1848, and opposed the Fugitive Slave Law, by 1860 he was a Unionist above all else, and had evidently reached the conclusion that the election of Douglas would be a blow at the Union. Accordingly he joined the Breckinridge-Lane party, presiding at its state convention in Syracuse, and being named as elector at large on its ticket. Efforts in New York at fusion of the Douglas, Bell and Breckinridge Democrats eventually led to the making of a slate of candidates for presidential elector containing representatives of all three

thoroughly crippled and humbled, and his name is William H. Seward.[89] I have read a little history in my time, both ancient and modern, but never have I fallen in with a match for Seward in duplicity, demagoguism [sic], heartlessness, low ambition, selfishness, and rotten corruption. Not a dime would he give to liberate every slave in our country; he would sell the whole Union for half the money Judas Iscariot received for betraying his master, and be —— to him. Marshal Rynders [90] is immaculate in comparison with Wm. H. Seward.

I am up to my eyes in business: have no time to say more.

Yours truly,<br>Geo. Wilkins Kendall

Rancho near New Braunfels,<br>Nov. 21, 1860.

My Dear Sir:—I have been so busy of late—so constantly on the move—that I have not had time to write you:—even

groups. Although Randall's name was at first included, it was eliminated by the Douglas group. His elimination was probably the result of the stand against fusion which he had taken in a letter to John Slidell which the New York *Evening Post* printed in part while the fusion slate was being considered. See DeAlva Stanwood Alexander, *A Political History of the State of New York* 3 vols. (New York, 1906–1909), II: chapters 20, 22 and 24; Sidney David Brummer, *Political History of New York State during the Period of the Civil War* (New York, 1911), chapters 2 and 3; New York *Evening Post*, 1 and 4 Oct. 1860.

[89] William Henry Seward (1801–1872) had aroused Southern resentment during the years before 1860 when, as a United States Senator from New York, he had taken a vigorous stand against slavery. It was Seward who, in 1858, called the struggle over slavery "an irrepressible conflict."

[90] Isaiah Rynders (1804–1885) was for many years a notorious figure in New York City politics. Polk, Pierce and Buchanan all appointed him to office, the latter making him United States Marshal for the Southern District of New York. Like Randall he was a member of the regular delegation from New York to the Democratic National Convention in 1860.

now I have but leisure to give you a short letter, to let you know that I am still in the land of the living. You, who have an old ancestral shelter over your head, cannot well imagine what a work it is to build up a home for a household—a household as numerous as your own. In two or three weeks I hope to move with my family to my sheep estancia near Böerne, and then I farther hope to have a breathing spell.

And about my household. I, too, have four children,[91] the eldest a girl, going on eleven, rather handsome than otherwise, and I think sprightly and promising. Of course she reads writes and speaks English and French, as both languages are used under my roof; more than this, she reads, writes and speaks German as well as either: I early took an inveterate dislike to that language, and finding that I could not master it determined that my children should. So, for the last eight years or more I have had a German nurse or governess in my house—for the last five a most excellent woman, finely educated—and you can readily imagine the advantages. My second child is a boy of eight years: he came with us to Texas in but indifferent health, and we hardly hoped to save him. But our mountain air and exercise have brought him out, and he is every day going stronger and stronger. He also reads, writes and speaks English, French and German, and his aunt, who came from France with us, thinks him intelligent. Our third child, a girl of seven years, is most unfortunately deaf—deaf from a fall while yet a child. She is a beautiful child, intelligent to a degree, and is now in Paris under the tuition of a gentleman who teaches

---

[91] Kendall's children:

Georgina de Valcourt (1850–1947). In 1873 she married Eugene J. Fellowes. She was much interested in the history of the Kendall family and in preserving family records.

William Henry (1852–1877). He eventually went to live with his father's sister in Vermont, where he died.

Caroline Louise (1853–1899).

Henry Fletcher (1855–1913). He graduated from West Point and had a military career, attaining the rank of major.

the dumb to speak.[92] She already reads, writes and speaks (French of course.) My partner, Mr. Holbrook,[93] visited her last summer, while in Paris, and informs me that she speaks well, and with a pleasant voice. She understands what you are saying by the motion of your lips, and answers as plainly as an ordinary child of her years. The gentleman who has charge of her education has another pupil, a deaf and dumb boy, son of a wealthy banker of Bordeaux, who also speaks fluently and well. Our little girl writes us a letter every two weeks: I need not say that no expense shall be spared to give her the best possible education. She has a married aunt in Paris who visits her often: this relieves us of much anxiety. Our youngest child, a boy of five and a half years, is as lusty and vigorous a specimen as you often meet. His mother thinks him the handsomest and smartest child now living, and I do not dispute her. Mrs. K. teaches our eldest girl music; before she is fifteen, if I live, I intend she shall learn Spanish and Italian thoroughly, knit my stockings, make and iron my shirts, and cook me a good breakfast, dinner and supper. My boys shall also learn all the languages, and be brought up as farmers and stock raisers if I can have my own way. My wife's mother is with us, a lady of finished education, and assists in teaching the children.[94] For the

---

[92] Pierre Auguste Houdin (1823?–1884). Teachers of deaf mutes differed in respect to methods and objectives of instruction. One school favored the use of sign language and the development of the ability to use sign language; a second school favored oral instruction and the teaching of pupils to speak and to read lips. Houdin, as the letter shows, belonged to the second school. In 1865 he memorialized the Imperial Academy of Medicine in a work entitled *La Parole Rendue Aux Sourds-Muets et L'Enseignment Des Sourds-Muets Par La Parole* (Paris, 1865).

[93] Alva Morris Holbrook (1808–1876), a native of Vermont, joined the *Picayune* partnership in 1839. At the end of his life he was the owner of the paper. His young widow, the poet "Pearl Rivers," proved herself an able newspaperwoman.

[94] Caroline Suzanne de Valcourt came from France at the end of 1858, bringing with her the three-and-a-half-year-old Henry Fletcher, who had been placed in her care three years before. A son, Adolphe,

three at home with us I have little solicitude: they shall be brought up to shift for themselves if need be. *I am working for our little girl in Paris:* she has been unfortunate, and shall want for nothing.[95]

I have now given you a hastily drawn picture of my little household: my brood is younger than yours, for I did not turn Benedict until I was nearly forty. When are you coming out to see us?

I read your letter in the Courier,[96] and thought it spicy and pointed to a degree. You did right not to publish a line

---

and a daughter, Henriette, also came to live in America. Mme. de Valcourt was drowned in 1873 in the Guadalupe River while returning from a visit to Henriette, who had married Cumming Evans. See Copeland, *Kendall.*

[95] Kendall invested wisely in real estate in San Antonio as a means of insuring the future of his handicapped daughter. The lots later became valuable. See Copeland, *Kendall,* 277.

[96] This was Randall's letter on fusion referred to in note 88. It was first printed in part in the New Orleans *Courier,* 20 September 1860. The *Courier* described it as "an exceedingly interesting private letter from one of the most intelligent and prominent leaders of the Democratic party in New York, to one of our friends in this city," giving the name neither of the writer nor of the recipient. The letter was dated 8 September 1860, at Cortland Village, N.Y. Although the letter was largely in opposition to fusion it contained a paragraph beginning thus: "But if the national sachems think differently of the policy of fusion—think it may elect Breckinridge—then at all events let us have the negotiations speedily brought to a termination, one way or another." The New York *Evening Post,* a Lincoln paper, reprinted the letter from the *Courier* on 1 October, omitting, however, the paragraph mentioned above. The *Evening Post* attributed the letter to Henry S. Randall but did not name the recipient. The context of this letter from Kendall suggests that Randall himself had informed Kendall that the fusion letter had been written to John Slidell.

John Slidell (1793–1871) went from New York to New Orleans in 1819, and by the time the *Picayune* was founded had become prominent in law and politics in Louisiana. He was a member of the U. S. House of Representatives from 1843 to 1845, and of the U. S. Senate from 1853 to 1861. He broke with the Buchanan administration in its final months, opposed Douglas at Charleston and like Randall supported Breckinridge and Lane in the election. During the war he was an agent in France for the Confederacy.

saying that it was not intended for the public eye: never do any thing of the kind: nine times out of ten explanations only make matters worse. A year or two since an article appeared in one of your Albany papers stating that your humble servant raised the capital on which he started the Picayune at a game of draw poker at the Astor House! The writer went on to say that I was working for Greeley at the time on the old New Yorker, gave the names of my companions, and every line of the article was true save the important item of winning the money. Some of my friends wished me to contradict it: not a bit of it—never. My intimate friends knew it was untrue, and for outsiders I do not care a fig. All the money I had when I started the Picayune was $75. My then partner, Lumsden—poor fellow! now dead and gone!—was about that much in debt. Such was our capital.[97]

Speaking of draw poker puts me in mind of your friend John Slidell, to whom you wrote confidentially. The fact is patent that there are worse brag and poker players than Slidell, and when he was appointed Minister to Mexico, in '46 I think, the following brick-bat, thrown by a Baltimore editor, struck him:

"The Hon. John Slidell has been appointed U.S. Minister to Mexico: if the Mexicans can win any thing off of him, it is more than we can do at home!"

Wasn't that a good one? I know that I laughed heartily over it at the time.

We have had no frost as yet: grass still fine, vegetation green, and plenty of tomatoes, snap beans, water melons, &c &c. on my table. How is it up your way? How's the sleighing?

I am off for my sheep estancia bright and early to-morrow

---

[97] Kendall's memory would appear to have failed him in one particular. He left New York City in 1833, while Greeley did not begin publishing the *New Yorker* until the next year. During 1833 Greeley was running a printing shop in New York.

morning, if it don't rain. On my return you shall hear again from

> Yours truly,
> Geo. Wilkins Kendall

Abraham Lincoln is President, and Wm. H. Seward is ——. We shall see what we shall see.

> Rancho near New Braunfels,
> Nov. 22, 1860.

My Dear Sir:—Here it comes, a soaking rain, and I must put off my trip to Post Oak Spring until it clears off. I had my horses fed at daylight this morning, brought out after our early breakfast, and was about to harness up, when down came the rain. Had it kept off ten minutes longer I would have been on the road; and as I never turned back in my life, when once started, I should have kept on. To-morrow I shall go up, rain or shine—at least unless it pours like cats and dogs.

A word about the buck No. 364, an animal which has roused a species of hornets' nest about our ears. After receiving your letter as to the whereabouts of the missing buck, with a copy of your letter to Mr. McCrearey, I wrote to the latter immediately. I did not hint that he had acted wrong in the matter, although the tone of my letter might have been cool. (But I am getting along too fast: it was immediately after receiving a letter from him, a day or two after yours came to hand, that I wrote.) I told him that No. 364 was just the buck I wanted, and that inasmuch as he had had the use of him this season, and had his strain of blood, while I had not used No. 183 at all, I thought it no more than fair he should be at the expense of exchanging. I told him, moreover, that the moment I missed No. 364 I wrote to you, and also to friends of mine on the Nueces to be on the lookout for him. This may have nettled Mr. McC. a little: I certainly did not insinuate that he had had any hand in the exchange, although it had been an unfortunate one for me, and had caused me much disappointment. I believe I have now given you the substance of what I wrote.

Well, the last mail brought me the enclosed letter from Mr. McCrearey, which, as he requests it, I send to you.[98] It speaks for itself. He also sent me the statement of the pilots who took his sheep off the vessel, which I also forward to you. All that the pilots say, all that the captain of the vessel says, all that Dr. McC. says, does not change my belief that

---

[98] McCrearey's letter, dated 15 Nov. 1860, gives the doctor's version of the affair: He was not on board the *Texana*, which, because it drew too much water, had to wait several days before entering Matagorda Bay. His sheep were taken out of the cages in which they had made the trip and brought ashore in a pilot boat. He put the sheep in a grazing yard, unaware that he had a sheep not belonging to him. Nearly a week later his son discovered that they had a ram numbered 364 instead of 183. Even then he did not know that No. 364 belonged to Kendall; he assumed that Randall had either made a mistake or had concluded to substitute No. 364 for No. 183.

The doctor denied that he had had sufficient use of the ram to warrant his being expected to pay the expenses of exchanging the rams. He offered to let Kendall sell No. 183 and send him the proceeds, and to take No. 364 to Indianola, transportation from Indianola to Kendall's place to be paid for by Kendall or Randall.

McCrearey's letter exhibits a dignified, intelligent and perturbed man. In view of his obvious innocence of any wrong doing in the matter, we can sympathize with him when he asks: "Now how have I offended, what have I done wrong. I ask you now Mr. Kendall as a gentleman of high character, and an honest man, if I ought to suffer morally or pecuniarily, for a thing of this kind that has been thrust upon me?" McCrearey's letter to Kendall, the certificate of the pilots, and letters from McCrearey to Randall are in the Randall Papers.

Although we have no letters from Randall on the matter, we may believe that the error was, as McCrearey suggested, made at Cortland. On 25 Sept. 1860 R. Herndon wrote to Randall from Texas notifying him of the arrival of some sheep which Randall had sent him. He noted that whereas he had ordered three ewes and one buck, he had received three bucks and one ewe. He added: "You have sent me somebody elses order & my order perhaps to somebody else. So you may be in two difficulties at once, But can make the mistake work very well & therefore have no complaint to utter." This letter is also in the Randall Papers.

either in New York or on the voyage, by accident or design, the bucks were changed. The captain of the vessel is said to be an excellent man, and very popular; yet the bucks might have been changed on shipboard as easily as an adroit gambler could slip a Jack, and he know nothing of it. From what I can learn Dr. McC. is a gentleman of good standing and veracity, and I have not the least suspicion that he was in any way privy to the exchange. Nor should I ever have had any suspicion of unfair play had not *the* buck happened to turn up missing—the largest buck and the heaviest shearer.

My impression, on learning that some 25 head of sheep had been taken off the vessel at Saluria, was this: that some man employed by the owner might have had this lot in charge—that he found No. 183 a pugnacious, quarrelsome animal, (he is a fighting devil,)—that No. 364 was quiet and docile—and that to save himself trouble he made the exchange, unbeknown to any one. Or he might have taken a fancy to No. 364, and swapped him surreptitiously. I could not believe that No. 364 jumped bodily, of his own accord, into Dr. McCreary's pen, and that No. 183, although a belligerent "cuss," jumped bodily out of his pen into mine: I don't believe it now.

If Dr. McCreary, on finding out that he had a buck he did not purchase in his lot, and if he knew that I had bucks on board the same vessel, which he certainly had a fair chance of knowing—had he then written me a line, all this fuss might have been avoided. I shall write to him, and tell him so. And I shall say to him, moreover, that I will pay the transportation on the buck No. 364 to my place here, and that I will either sell his buck, No. 183, to one of my neighbors, or send him back at his cost. It will cost some $10; but not for ten times $10 would I have had all this fuss, which is annoying all round.

I had hardly commenced this before company came in, and I have written in a great hurry. I shall say to Dr. McCreary, in my letter to him, that I am satisfied with his explanation. He was evidently a little sore when he wrote to me; yet it was only by forcing a construction on what I wrote to him that he could for a moment think I considered him as in any way guilty of wrong.

In haste,

Truly your friend,
Geo. Wilkins Kendall

I will say more about this matter in my next.

Rancho near New Braunfels,
Tuesday, Dec. 4, 1860.

My Dear Sir:—Just as I was starting for my Sheep Estancia, on Friday last, I received two letters from you: I returned last evening, and this morning, although I have rummaged about and given my writing desk a general jail delivery, I can find but one, dated Nov. 10, and giving me a plan for a prison for "Old Honest" next year. The other was a short note, and its contents I have now forgotten.

If ever any poor devil's time and brains were taxed, I am that unfortunate individual just now: Caleb Quotem [99] had not half the work on his hands. In addition to my ordinary work, I have a lot of masons, carpenters, painters and what not to watch, and besides, new comers, prospecting for locations, are flocking in, and all fall upon me as a species of directory and general intelligence office. I would like to play Robinson Crusoe for a month, and not see a single soul—not even a man Friday. Pope had not half the trouble when he wrote his Epistle to Dr. Arbuthnot.[100] At times I wish that my Estancia was in the Llano Estacado [101]—a hundred miles

[99] A jack-of-all-trades in George Colman the Younger's *The Review, or The Wags of Windsor* (1798).

[100] In his *Epistle to Dr. Arbuthnot* (1735), Alexander Pope describes the importunities and criticisms which beset him.

[101] The Staked Plain, a vast high plateau extending from the middle of western Texas over most of the Panhandle of Texas and eastern New Mexico.

from nowhere. Job, had he been in my fix, would have cursed and swore worse than one of Uncle Toby's troopers in Flanders: [102] you cannot begin to imagine how I am beset and encompassed about. When I move up to my sheep place I believe I will stick up a huge sign, "No admittance even on business," for a month or such matter, and see if it will bring a short respite. But a truce to complaining: I presume I shall worry through.

It makes me shiver to hear you describe all the preparations you were making, a month ago, for winter. Five of my flocks, and among them the two breeding ewe flocks of over 1000 each, must go through the winter again without other shelter than rails. They are all in the finest possible order, and I am in hopes will be able to "tough it out" without suffering or loss. All are young, well coated with wool, and so far we have an abundance of excellent grass.

"Old Honest" has got over his tantrums, and is now as well behaved and as quiet as a Quaker. I'll have a place for him next year, if I live, which will hold him, if I have to put him in my cellar. He conquered every thing this year.

I have sent McCrearey a long letter, which I trust will operate as a slave to his wounded feelings. I quietly told him one thing, however, which he may not like: I said something like this:—"Living within a stone's throw of Indianola, you might easily have ascertained to whom the six bucks, landed at that place, belonged, after you found that you had an animal that did not belong to you, and then a single line to me would have explained every thing, and set all right." This he should have done, and I could not help telling him so. I am told that he is a gentleman of good standing, and that his wife is a most amiable and ladylike woman. But until all this fuss occurred, I had not the most remote idea there was such an individual in existence as Dr. J. K. McCrearey. He has asked me to sell his buck if possible, and I

---

[102] " 'Our armies swore terribly in Flanders,' cried my Uncle Toby, 'but nothing to this.' "—Sterne, *Tristam Shandy*, Bk. III, Ch. 11.

have written him that I would send for mine by first mule team going down. Saluria is within sight almost of Indianola.

Even yet I have not gold [got] hold of the Texas Almanac for '61, but am looking for it every day. Your article will be of great advantage to new beginners in sheep raising, provided they will only follow its instructions. But will they do it? There's the question. A new comer the other day, who knows just about as much of sheep raising as a cow does of playing on the trombone, told me that my system was all wrong: that he would beat me out of the field in five years. If he does not come to grief before the present winter is over, I am very much mistaken. "Your sheep go out too far every day," said he, "and are over-worked." "Tie yours to a bush," retorted I, "with three feet slack of rope, and then they will not die from travel." You have little idea of the stupidity which obtains among new comers hereabouts.

You ask me about "Secession," and whether I have observed enough to form an opinion whether it is likely to take place or not? I have only ten minutes to answer your question.

We have never had but two administrations—that of the older Adams, who *wanted* the people to give him power, and that of Gen. Jackson, who *took* the power whether the people would or not. The weakest of all our Presidents is James Buchanan: he has looked timidly on while half a dozen Northern States, your own among the number, have virtually nullified—have openly trampled the Constitution under foot by their action in relation to the Fugitive Slave Law. The people of the South are not so particularly stupid as not to know this: they would be more than human were they to look on patiently and see the additional encroachments upon their rights made by your Northern Druses—preachers of "higher law," and instigators of "irrepressible conflicts." I am not a Secession man myself—I am still for attempting to hold our Union together. But I would have the entire South hold counsel together, decide upon a firm and dignified plan of action, give the North an unmistakable ultimatum, and in the meantime prepare for the worst. But I am all the time doubtful whether there is good faith enough left in the North to do justice to the South. If you in the North openly nullify by setting aside a plain and palpable provision of the Constitution,[103] any State in the South has an undoubted right to secede or nullify. If the Union is once broken, it will fly into fragments, and the verdict will be that it died of its own weakness—from its utter inability to resist the onslaughts of mad fanaticism, and protect the just rights of its weaker members. Oh! that Andrew Jackson was now at the head of the Government, and that he had a standing army of 200,000 regular soldiers to back him. Our country is far better able to support a standing army of soldiers of the above number, than the present standing army of regular office holders and office–seekers— 500,000, at least.

As for Texas, so far as I can gather, she is ripe and ready for Secession, and if the Union is to be split I believe that our best course will be to "go it alone." [104]

I have no time to "define my position" further: could I have an evening with you, you would find me as near right as most men, and a thorough Union man and American.

In haste,

Your friend truly,
Geo. Wilkins Kendall

I am off again for Post Oak Spring: will write on return.

---

[103] The Constitution provides for the return of persons "held to service or labor" in one state who flee to another state. The Fugitive Slave Law of 1850, based on this provision, was openly flouted in the North.

[104] A secession ordinance was passed 1 Feb. 1861 by a convention meeting in Austin. Submitted to the people for ratification, it was approved 23 Feb. On 2 March Texas became a member of the Confederate States of America.

# THE LETTERS OF 1867

Boerne, Kendall Co. Texas,[1]
April 7, '67

My Dear Randall:–If ever you have read the Pilgrim's Progress, let me tell you that Bunyan's hero had a macadamized road with good stopping-places every night, in comparison with what I have encountered since I last wrote: his trials and tribulations were as cakes and gingerbread compared with mine. I reached my old and pleasant (once pleasant) home three weeks ago to find my house deserted and in dire disorder, dreary and dilapidated;[2] my flocks doing badly; the season a month later than usual; no grass when the prairies should have been as green as wheat fields: –in short, every thing demoralized. Lambs commenced coming on 12th March, at the end of a warm and growing spell: on 13th the worst sleet storm ever experienced thus late set in: away went grass: ewes had no milk: consequences you can easily judge. I hoped to raise 1800 lambs this season at least—if 500 rub through I shall be satisfied. Most of my neighbors—all with large flocks—served in the same way. Weather continually and persistently cold, raw, and backward. How with you? In the older Southern States:–floods here, freshets there, and late frosts every where. Never have I experienced the like—never care about witnessing a repetition. My flocks, Randall, have never had so slim a chance to get safely over or through the winter, as the young men who have had charge of them have over-cropped themselves—

---

[1] The Kendall family moved to their new home at Post Oak Ranch in February, 1861. The following year a new county was created from Kerr and Blanco counties and named for its distinguished sheepman. Boerne became the county seat.

[2] "When I arrived here I found my house in a miserable condition, dirty and dilapidated. I felt, as if I wished to run from it as far as I could, and then lie down under a tree and die, but my better sense came to me and I went to work cleaning. I sent for a mason, and did not rest until our humble little house was clean."—Adeline de V. Kendall to Mrs. Henry S. Randall, Post Oak Spring, 20 Sept. 1867, New-York Historical Society.

have had two lots of sheep besides mine to attend to. The consequences you can easily imagine—more especially when you take into consideration the persistently cold snap which has followed the awful sleet storm of the 13th ult. Yet I am as hopeful of the future as ever I was—I know what I have done in Texas in years past, especially before the war, and unless Providence has entirely changed the seasons, I am sanguine I can effect the same results again. Why not?

Do you get the Weekly Picayune regularly? Let me know. I told them to send it to you. If you receive it, you will have seen occasional signs that I have not been altogether idle. I shall keep up a regular correspondence on the sheep question hereafter, shall tell the truth if I know how, and wish you to read what I write.

I commence this after a hard day's work: will add to and finish it whenever I can steal idle moments. Oh! for a little leisure.

*April 9.*–A warm and growing day at last—the first in nearly a month. But so treacherous has been the weather this month, and nearly all the last, that I will not crow: our old ewes are picking up a little, and the curled up lambs, what are left of them, are straightening out:–but we are not yet out of the woods.

A letter from McKenzie at last.[3] It is dated Oakville, Live Oak County, Feb. 5, '67, and is over two months old. As I believe there is no mail to his section I presume he handed it to some traveller, who "pocket-vetoed" its transmission. I am sorry to say that the beautiful ewe you presented me died shortly after arrival, but both the bucks are alive, and doing finely. Let me give you an extract from my old head shepherd's letter:–

"The two bucks are a perfect show: I have one man to take care of them, and nothing else. As soon as I get a little

---

[3] John McKenzie, described by Randall as "one of the best flock-masters in Texas," in the *Texas Almanac for 1868*, 169. He was Kendall's head shepherd for a time during the war.

time I will take yours up to you. I shall write to Mr. Randall in a few days, and send him his money. *I never saw such a buck as mine in God's world.* [Italics mine, but language McK's.] Capt. King [4] and Bryden [5] are going to get bucks from the North—I told them to send to Randall."

Such is the extract. You have a friend in McKenzie, Randall, from this out, and one who will be of service to you if all goes well with him. He is a watchful, hard-working, plain, sturdy, honest Scotchman, and driving withal. I presume he is now out on the prairies away from the post roads, post offices and settlements, hunting grass, and he'll find it.

*April 11.*–I will finish and send of[f] this poor apology for a letter—or such a letter as I owe you—promising a better next time.

You should know that I remained in N.O. two months longer than I expected, waiting for French "help" which my brother-in-law hired for me in Cherbourg. But I was not idle; on the contrary, I was busy fifteen hours a day on the Picayune—had not a moment to call my own. One of my partners was detained by illness in Mississippi: off-and-on one of the others was down all winter. So that I was compelled to carry a double or triple load.[6] Had no time to write to you or any one else, but my wife and Georgina kept you apprised, I believe, that I was in the land of the living. At all events I told them they must keep up my correspondence.

Well, at length my "help" arrived—a man, his wife, boy of 15, and girl of 13 years of age; I started them at once up here; followed by stage a few days after: reached here one day in advance; have set them all to work. The man is an excellent gardener and farm hand; the woman a good cook, the boy as spry as a weasle at "chores" and the like, and the

---

[4] Richard King (1825–1885), the founder of the famous King Ranch between the Nueces and the Rio Grande rivers.

[5] A rancher, and also an associate and foreman of King.

[6] For an account of Kendall's busy winter in New Orleans, see Copeland, *Kendall*, 311–16.

girl quick to learn household work. We are well off in this kind of help, and they are bound to us for five years. So soon as the time of the men who have leased my sheep expires (1st of Sept. coming,) I shall give my flocks a regular over-hauling—cull out and sell all the old sheep—and enter next winter with about 3500 active producing animals only. Now, if you know or can hear of a thoroughly active shepherd— one who understands his business all the way through—and is anxious to better his situation, ask him to open a cor-respondence with me. My sheep *shall* be cared for.

We have our youngest daughter with us: my two boys are at the Episcopal school in San Antonio: Georgina remains at Miss Hull's Seminary in N.O. until May to graduate: my wife joins in kindest regards to your entire household: and I am

> Your friend
> Geo. Wilkins Kendall

> Boerne, Kendall county
> May 6, 1867.

Your long and welcome letter of the 9th ult. my dear Ran-dall, I found here on returning from New Braunfels, which is thirty-five miles north of east of my Post Oak Spring place. They would make me a director in a new woolen factory [7] about to be started at that city, and I lost some four days from my work helping organize. A cotton factory is already in "full blast" at New Braunfels,[8] and you may make a

---

[7] The New Braunfels Woolen Manufacturing Company, which com-menced operations in June, 1868. It was soon turning out an average of about 200 yards of tweeds or jeans and 40 pairs of blankets daily. Texans looked forward to New Braunfels' becoming "the Lowell of Texas." See the *Texas Almanac for 1869*, 150.

[8] The Comal Manufacturing Company, established to produce cot-ton goods, began operations in 1865, using the power furnished by the Comal River.

memorandum that that city will be the largest manufacturing place in the South, some day or other. It has *all* the advantages.

I had already read the tariff article in the Cortland Gazette. Every wool grower in general, and in Texas in particular, owes you an eternal debt of gratitude, Randall, for your exertions, and the Texas people shall be made to know and appreciate it all in good time—your labors for their protection.[9]

So, at last you have a letter from McKenzie. He is a queer chap, is McK., eccentric and full of prejudices, but strictly honest withal, a good judge of sheep, industrious, watchful —a sturdy Scotchman. I have not heard from him since February, but look for him up here after shearing time. What kind of luck he has had this spring—this unprecedently cold and backward spring—I have not learned. I presume he has suffered along with the rest of us, but he has lost nothing by neglect, by default, or want of early and late striving. That storm of 13th March, with the raw Norther which continued all through lambing time, spunged [sic] out all our profits for '67. Yet I am confident McKenzie lost less than any one.

The men who have my sheep until weaning time in August have now been shearing a week—four weeks behind the usual time of commencing. Two days before they began a heavy, beating, drenching shower as completely washed the sheep as though they had been put through the Northern process. The wool is clean, dry, and of course light. I shall insist upon Field's setting it down as *washed* wool, at least the first part of the clip shorn, because *it is washed* wool.[10]

---

[9] See pages 23–24, above.

[10] Northern sheep raisers often washed their sheep in a stream a little before shearing time, and sold their fleeces as "washed wool." Buyers commonly deducted one-third of the weight of an unwashed fleece to determine its "washed wool" equivalent. This practice, Henry Randall argued, was "to offer a premium for neglect," since a conscientious farmer who looked after his sheep might have wool as clean

And should Field put it into market in *Ohio* wool bags, now? As Texas wool, although fine, it will never get its deserving. They will be nearly or quite a month running the flocks through, although they have nine shearers and three to tie up and bale. The latter portion of the clip may be more oily, dirty, and heavier of a certainty. My sheep have improved wonderfully since I last wrote, and the lambs as well: I wish you could see them.

You speak of your desire to send out a lot of fine ewes to sell. As I shall not this year have the means to purchase— shall have hard scratching to get through until the spring of '68 and make all the improvements I require—I will tell you what I can do for you. I have already asked you to look and enquire about for a thoroughly competent man to take charge of my flocks at weaning time, say about middle of August. Now, if you find such a man, and feel disposed to send out 20, 30, or even 50 ewes by him, starting say in July from New York for Indianola, I will take charge of these sheep for you, sell them for you, and keep them for you free of charge until sold. On arrival, I will blow—not a single trumpet, but a band of trumpets for you—that is, I will get every editor in the State "tooting" in behalf of the sheep. If they cannot be sold this fall, at least all of them, I will keep them in my pastures and rye fields for you through the winter, and raise lambs from them. This would involve the necessity of your sending out a buck with them, for service in October. And all this I will do willingly, cheerfully, and with no other reward than the gratification it would afford me to repay the many obligations I owe you. This influence I have: every editor in Texas is friendly to me, and will publish any thing I may wish to insert free of charge.

I cannot say how it will be this year about selling sheep. I know that most of the wool producers are cramped. It is

without washing as another farmer had after the sheep had been run through a stream. The question of "washed" versus "unwashed" wool aroused much discussion and bitterness.

costing us all a deal to get rid of the scab, which we are
rapidly doing. Our cold and most inclement spring has been
terribly against us. I do not say that many are disheartened
—I know that I am more sanguine than ever. But you can
readily imagine that the flockmasters have suffered severely,
and it will take a year to get over the evil effects of the
present spring. Most anxious am I that you should retain
and increase your Texas business, and so you shall; but do
not look for too much the present unpropitious season. I
have not as many lambs by 800 or 1000 as I should have:
this tells a story, *the* story. I fancy I am as well off as many
of the heavy sheep owners.

A man named M. M. Barber writes me from Austin that
he has recently arrived there with a lot of American Merino
bucks from Addison county, Vt.,[11] and modestly requests me
to go over and see them! I have respectfully declined. From
the tone of his letter I think he is making out but poorly:
he says that if I will visit him he will offer me cheap bar-
gains.

As to the future of Texas as the great wool producing region of the Union [12] I am even more sanguine than ever—the
ultimate effects of the new tariff, which may be modified but
which I doubt will ever be materially changed, will be all
in our favor. Every mail brings me letters from anxious enquirers in the old States about the prices of land, sheep,
prospects, and what not. Strangers are also flocking in, and
when they find they are not to be bowie-knifed on sight,
thousands more will be coming. Of this I am sure.

You can think over the matter of sending out sheep for

---

[11] Addison County was a leading sheep breeding center. Middle-
bury, in this county, was the home of Edwin Hammond, probably the
best known merino breeder of his day.

[12] Kendall's optimism concerning the future of Texas as a wool
producer was justified. In January, 1955, Texas easily outranked all the
other states in the number of sheep and lambs, having 5,331,000 of
the country's 30,931,000. Wyoming, her nearest competitor, had
2,132,000.

me to sell or manage for you. If you procure me such a man as I require—one who understands sheep, and who is industrious, active and watchful, and give a small flock to his charge, it would insure their safe transit. And I repeat that any I cannot sell this fall will be more valuable next year on account of the lambs they will bring you.

I have been working hard, very hard, since my return, setting things to rights and putting my house in order. I have a deal yet to do. I intend, as you say is best, to keep my sheep, instead of leasing them out on shares; but as I have fences to make, pastures to enlarge, crops "to pitch" in the fall, orchards to set out, and what not to engage my attention, I wish to hire some trusty man, of *practical* knowledge, to watch both sheep and shepherds. Out of your legion of friends and acquaintances I trust you can find some one who will point out such a man. I will pay such a man liberally the first year, and, if he suits me, give him a flock of ewes the second on shares. A *practical* man.

*May 8.*–A regular cold, raw, whistling Norther yesterday, with slight frost! This beats all. To-day a blustering South wind, dry and searching. So tell me what kind of a season you are having. Hereaways, every thing seems completely turned around.

I send you a Galveston News with a sheep article. Bad as the season has been, where flocks have been managed by thoroughly *practical* and industrious men they have done tolerably well in Texas: your *theoreticals* only have lost to any great extent.

There are two callings, friend Randall, in this world, which imperatively demand *practical* knowledge, viz:–navigation and sheep raising. In fair weather all goes well with theory, but when storms and foul weather come, then a practical man is needed at the helm. You may make a voyage to Calcutta and back, and watch every rope and every movement on board from daylight until dark, and be up of nights as well: you may think you thoroughly understand naviga-

tion: but the first gale, cyclone, hurricane or tempest that comes, where are you? Gone! So with navigating a flock of sheep through the voyage of a year in Texas: you must serve your time "after a flock," as "before the mast," in order to insure a safe termination. The men who have had charge of my flocks are honest and conscientious, watchful enough, and anxious to achieve success. But they came into business "through the cabin windows"—never spent a night in the "forecastle." You understand. They are sufferers—so am I. But I am tiring your patience, trespassing on your time, and will hold up. An active, watchful, industrious, *practical* man is what I want.

My wife, who is hard at work making trousers and jackets for our boys now at Bishop Gregg's school in San Antonio,[13] says, with her compliments, that she will write you so soon as she can steal time. When I reached San Antonio I found my boys poring over Latin grammars: these I told them to throw out of the window, and buy Spanish grammars instead. The useful before the ornamental in hard times. Georgina has thoroughly mastered Spanish the past winter: next month she comes home to help her mother: housework of every kind she now understands tolerably well—her knowledge will be thoroughly *practical* if we both live until next winter.

You want that group of photographs eh? That ancient phiz of mine, you should know, and which you think has a subdued look, was fastened on the plate just as I was on the point of sneezing! I had a horrible head cold at the time. The women folks—wife and two daughters—are all well enough I believe. We have a front lot still open in our album for your household, and all shall be bestowed in good company.

---

[13] Alexander Gregg (1819–1893), the first Episcopal bishop of Texas, visited the North in 1866 to raise money for church purposes, including education. He secured money to establish a boys' school in San Antonio; the Kendall boys must have been among the first students of the school. See Wilson Gregg, *Alexander Gregg, First Bishop of Texas* (Sewanee, [1912]), 90.

The shearers just tell me they have not twine enough to go through. I have a little business in San Antonio, and shall run down to-morrow morning after the twine and other articles. Shall post this letter there, and will leave it open to add a line.

*San Antonio, May 9.*–Came down from my rancho this forenoon, in an open wagon, and in face of a hot, boiling sun. Am stewed and burnt. Dry times here in San Antonio. Much more rain up my way in the mountains. Intended to have that group of photographs, with subdued old Pater Familias on point of sneezing, but forgot it.–Will send.

With kindest regards to your entire household I am   Ever your friend

Geo. Wilkins Kendall

Boerne, Kendall county
June 23, '67.

My Dear Randall:–Your prompt favor of the 31st ult. came to hand by last mail, and I answer by return post.

I note all you say about a shepherd, and on a separate page will describe such a man as I require and what I am willing to pay for services. I would prefer a trusty, sturdy, plain honest Scotchman: but to any man I can insure the basis of a fortune who in the first place understands *how* to watch and work, and in the second *will* watch and work.

In '52 on the recommendation of Gov. Paine [14] and Judge Follett of Vermont, I hired a man to come out and take charge of my sheep. He was highly commended, was smart enough, but was simply not worth shucks—was worse. He was with me five years, (I in France most of the time,) and during that period I did not "hold my own." It was only in

---

[14] Charles Paine (1799–1853), a prominent Vermont businessman, was governor of his state in 1842 and 1843. He was a woolen manufacturer, a stock breeder and a railroad promoter.

'56, when I came here myself and took hold of the helm, that matters mended, and from '56 to '61 no man ever made money so fast as I did at wool growing: never. Neither ancient nor modern history, Heathen nor Scandinavian mythology, mention any instance of the kind. And "it is in the cards," to win again if I can only get my flocks well managed.

Of course the war, the killing my shepherds, the advent of the scab, and adventitious circumstances following over which I had no control, literally tore me to pieces. But I have some five thousand head of high bred sheep to commence with anew, and repeat that all I now need is some honest, practical, working man, and watchful, to assist me. Much of my own time is taken up with farm work, correspondence, and what not: an eye on the sheep and especially on the shepherds, is at all times a matter of first necessity. It is in my power to make the fortune of any man who will do his duty, first understanding his duty. Am I sufficiently explicit?

On another page I will endeavor to explain what or who I want—something you can show. He must be a single man, and I say again that I prefer a Scotchman. According to your letter, the Vermonters are too exalted in their notions: I cannot afford to furnish a band of music. Scattered over the North I have no doubt there are many Scotchmen, and Americans too, who would suit me to a T. Now for a description. (Read memorandum.)

There, my good friend Randall, is that plain enough, presuming you have just read the memorandum? I have promised the man $500 in hard or sure-enough money—not so-called dollars of the paper persuasion [15]—with a promise of a

---

[15] A reference to the legal tender paper notes (greenbacks) which the U. S. government issued during the Civil War, and which were not redeemable in gold until 1879. They fluctuated greatly in relation to gold; during 1867 a greenback dollar was worth between 69.7 and 74.3 cents in gold.

present at the end of the year if he serves faithfully and successfully. You *might*, with all your care and caution, send me out a man who would not suit. And then any man who comes out should understand that he has a chance to finish his apprenticeship: for however well he may understand the business North, he has much to learn and much to unlearn in Texas. This is a fact and should be a consideration. But enough of this.

Now about your sending out some sheep. I can only reiterate my promise to take the best care of them possible, to advertise them, to show them, and to blow as well. I am now more hopeful that they can be sold than I was. Our crops in Western Texas this year are simply immense, grass was never so fine, stock of all kinds never so fat and healthy, nor general prospects of the farming and stock-raising people so flattering. Any one now hunting a home among us will be sure to cast anchor in our midst: cannot help himself, or themselves. I do not say that the stalks in my corn field are as high as those magnificent trees in front of your family mansion; but they are taller than any you ever saw, and eared out fully. And sweet potatoes! Well, it would gladden your eyes to see my garden and fields. And such a season and such crops will bring emigration, and emigration will bring money, and there will be a prosperity among us I did not dream of during our cold, backward, cheerless spring. But if you send out sheep you must send a man with them to care for them on the way.

I note all you say about the woolen factory business. You are all right so far as regards such matters North, but not in New Braunfels. There they have the wool at the doors; there they have an abundance of labor; there operatives can be fed and sheltered at ¼th the cost North; there they have a constant demand for coarse manufactured stuffs; there they obtain 25 per cent. more, and get their raw material 25 per cent cheaper than with you; and there the concern will

be managed by prudent, plodding Germans and sharp Yankees. It will win heavily. I am not interested to any great extent: wish I was.

Have not seen McKenzie as yet, but am looking for him every day. Presume he has been busy shearing.

Georgina graduated in N.O. with all the honors; is on the way home, and we are anxiously looking for her. She goes to work with the rest of us.[16]

This is Sunday: I have hurriedly written this, and must now as hurriedly commence a letter for the Pic. and other correspondence. Have not a moment to call my own. My French "help" proves help indeed.

On reading over "Memorandum" have added a Note.

Mrs. K. joins in kindest regards to you all, and I am, Ever your friend,

Geo. Wilkins Kendall

*Memorandum.*

I wish to procure the services of a man who thoroughly understands the management of sheep—who has been

---

[16] "Georgina came home in July, she is a great help to me, though I think that she is too young to leave her studies entirely, and I want her to practice her music every day. There is so much fun and life in her that she animates our home."—Adeline de V. Kendall to Mrs. Henry S. Randall, Post Oak Spring, 20 Sept. 1867, New-York Historical Society.

A traveler in Texas in the early 1870's saw Georgina ride up on horseback to get her mail at a little post office, and was quite taken by her. He wrote: "She is a handsome young girl . . . has an exceedingly intelligent face and a splendid figure. She was educated partly in New Orleans and partly in France, but mainly under the tuition of her father, and not less accomplished mother. She is noted for her horsemanship, understands the dairy, and is, perhaps, better qualified to govern a sheep rancho than any man in Texas. She is often seen mounted upon her spirited steed, 'a Tartar of the Ukarine breed,' driving in the herds of horses or cattle. I doubt not she would adorn the most elegant salons of the world, as she does her own beautiful prairie home."—From the Columbus, Ohio, *Times*, reprinted in the *Texas Almanac for 1872, and Emigrant's Guide to Texas*, 76–77.

brought up among them—who is acquainted with their ailments, and how to cure. I never have any foot rot or liver rot—no disease except the scab. My sheep now have it slightly, but I have the tobacco,[17] and will cure it within the coming six weeks, and drive the flocks to a new range. I *will* eradicate this pest.

A man coming here from the North or from Scotland, no matter how great his experience at home, has much to learn and much to unlearn: our management of sheep in Texas is necessarily different. I will pay any man Mr. Randall may recommend $500 in gold or silver for his first year's services: if he manages the flocks well, and we can agree, I will increase his wages to $1000 in gold or silver the second year, or give him an interest in the sheep which will enable him to make or realize a larger sum. I will almost insure the fortune of any honest, prudent, practical, pains-taking, watchful and industrious man—(all depends on the man himself) —within five years.

I have 5000 acres of land of my own, and command some 10,000 acres adjoining over which my flocks roam at will. At my homestead I have a cool and ever-running spring of delicious water, clear as crystal, which forms a large brook at the fountain head. The region is hilly or mountainous, the climate temperate for the latitude, and the health of every soul in the place is and has been invariably good: during twelve years a doctor has never visited me professionally. We are free from mosquitoes, Spanish moss, and malarias of all kinds.

---

[17] In his article, "Sheep-Raising in Texas," in the *Texas Almanac for 1867*, 217–19, Kendall gave detailed directions for preparing and using a sheep dip. During the war with wool prices low and tobacco prices high, it had been necessary to use an ineffective dip made with lye, lime and sulphur. "But," wrote Kendall, "tobacco is a sovereign cure, and with this at present prices, there is no more excuse for a man neglecting to cure his sheep than there would be if he saw the roof of his house on fire and poured the bucket of water in his hand out the window instead of on the blaze."

After weaning time in August, and after culling and selling old ewes, bucks and wethers, I shall probably commence the coming year with 4000 head of strong sheep of
which 1500 will be breeding ewes—perhaps 1650 or 1700:
I cannot now exactly say what number. For lame, disabled
or ailing sheep I shall have pastures and fields of rye and
wheat for winter. Here we can feed our rye and wheat
from 1st November to 1st April—always green and growing.

In conclusion I will state, that should any one come out,
and work faithfully and successfully the first year, I will
cheerfully make him a present at the end over and above the
$500. He must be a single man.

Geo. Wilkins Kendall

Boerne, Kendall co. Texas, June 24, 1867.

### Note to Memorandum

Should any shepherd come out, on Mr. Randall's recommendation, he will have a comfortable room to himself in
the same building with the under shepherds; yet he need not
expect all the conveniences and luxuries in a new which he
has been used to in an old settled country. His business will
be to look over the flocks daily to see if any thing is amiss;
to ride over the range on horseback to see that the under
shepherds attend to their flocks; to be active and vigilant
at all seasons, and especially at lambing time; to "camp out"
with the main ewe flocks perhaps for a couple of months
during winter, so as to save the home grass at or for lambing
time:–in short, to do all and every thing which may secure
the well-being of the flocks. The labor is nothing provided
any one has the understanding as to how to apply it, and the
will to bestow it faithfully.

I have said that I wanted a single man. I would have no
objection to a married man, but would not deem it best to
bring out a wife and children at the outset. Better come

out alone, and then if every thing goes well, and matters look promising, the family can be sent for.[18]

G.W.K.

---

[18] "My husband hired, three weeks ago, a Scotchman as head shepherd. I think that he is a good man."—Mrs. Kendall to A. M. Holbrook, 27 Oct. 1867; a printed copy of this letter, which was written six days after Kendall's death, is in the New-York Historical Society.

# INDEX